ASPIRE
SUCCEED
PROGRESS

AF774549

Cambridge Lower Secondary Complete Biology

Ann Fullick

Second Edition

TEACHER HANDBOOK

OXFORD
UNIVERSITY PRESS

OXFORD
UNIVERSITY PRESS

Great Clarendon Street, Oxford, OX2 6DP, United Kingdom

Oxford University Press is a department of the University of Oxford.

It furthers the University's objective of excellence in research, scholarship, and education by publishing worldwide. Oxford is a registered trade mark of Oxford University Press in the UK and in certain other countries

First published in 2021

British Library Cataloguing in Publication Data
Data available

978-1-38-201842-5

11

Paper used in the production of this book is a natural, recyclable product made from wood grown in sustainable forests.

The manufacturing process conforms to the environmental regulations of the country of origin.

Printed and bound by CPI Group (UK) Ltd, Croydon, CR0 4YY

Acknowledgements
The publisher would like to thank the following for permissions to use copyright material:

Cover: htu/Getty Images.

Artwork by Q2A Media, Integra Software Services, Erwin Haya, Barking Dog Art, and OUP.

Every effort has been made to contact copyright holders of material reproduced in this book. Any omissions will be rectified in subsequent printings if notice is given to the publisher.

This Teacher's Handbook refers to the Cambridge Lower Secondary Science (0893) Syllabus published by Cambridge Assessment International Education.

Welcome to your **Cambridge Lower Secondary Complete Biology** Teacher Handbook. This Teacher Handbook has been written to provide classroom support and teaching materials across all three stages of Cambridge Checkpoint.

Your Teacher Handbook includes a book of lesson plans as well as answers to all of the Student Book questions for your reference at any time. The Workbook answers can be found in the Workbook itself. The accompanying Kerboodle course includes a wide variety of additional resources to support you and your students in the classroom.

Using your book

This book contains suggested lesson plans and answers to all of the questions in the Student Book. There is also information about students' prior knowledge.

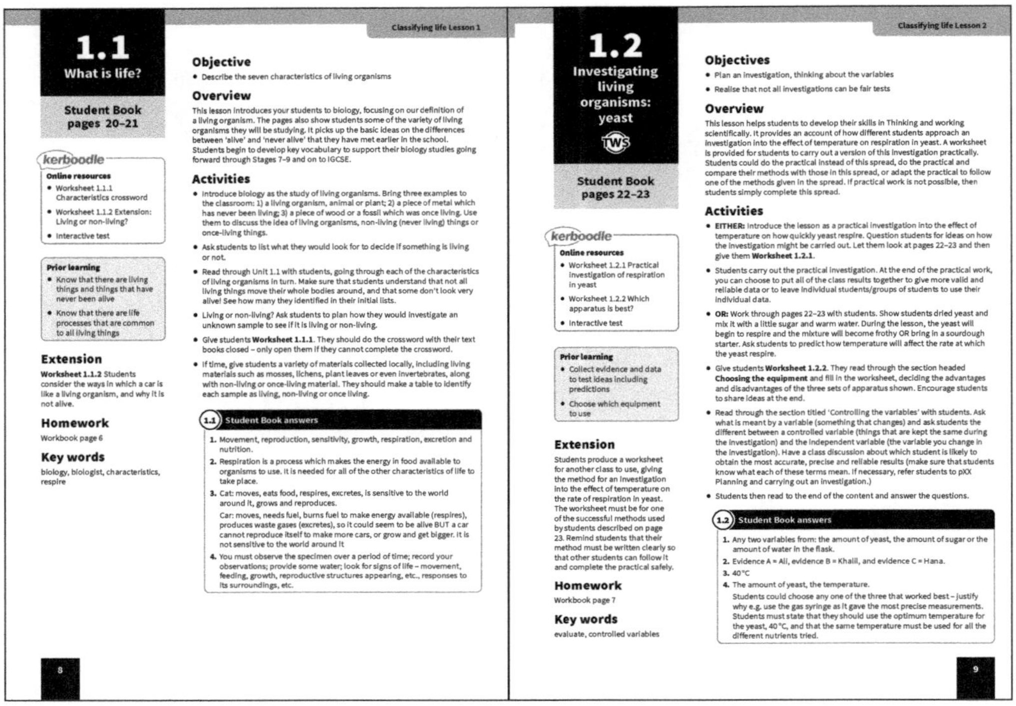
Classifying life Lesson 1

1.1 What is life?

Student Book pages 20–21

kerboodle
Online resources
- Worksheet 1.1.1 Characteristics crossword
- Worksheet 1.1.2 Extension: Living or non-living?
- Interactive test

Prior learning
- Know that there are living things and things that have never been alive
- Know that there are life processes that are common to all living things

Extension
Worksheet 1.1.2 Students consider the ways in which a car is like a living organism, and why it is not alive.

Homework
Workbook page 6

Key words
biology, biologist, characteristics, respire

Objective
- Describe the seven characteristics of living organisms

Overview
This lesson introduces your students to biology, focusing on our definition of a living organism. The pages also show students some of the variety of living organisms they will be studying. It picks up the basic ideas on the differences between 'alive' and 'never alive' that they have met earlier in the school. Students begin to develop key vocabulary to support their biology studies going forward through Stages 7–9 and on to IGCSE.

Activities
- Introduce biology as the study of living organisms. Bring three examples to the classroom: 1) a living organism, animal or plant; 2) a piece of metal which has never been living; 3) a piece of wood or a fossil which was once living. Use them to discuss the idea of living organisms, non-living (never living) things or once-living things.
- Ask students to list what they would look for to decide if something is living or not.
- Read through Unit 1.1 with students, going through each of the characteristics of living organisms in turn. Make sure that students understand that not all living things move their whole bodies around, and that some don't look very alive! See how many they identified in their initial lists.
- Living or non-living? Ask students to plan how they would investigate an unknown sample to see if it is living or non-living.
- Give students **Worksheet 1.1.1**. They should do the crossword with their text books closed – only open them if they cannot complete the crossword.
- If time, give students a variety of materials collected locally, including living materials such as mosses, lichens, plant leaves or even invertebrates, along with non-living or once-living material. They should make a table to identify each sample as living, non-living or once living.

1.1 Student Book answers

1. Movement, reproduction, sensitivity, growth, respiration, excretion and nutrition.
2. Respiration is a process which makes the energy in food available to organisms to use. It is needed for all of the other characteristics of life to take place.
3. Cat: moves, eats food, respires, excretes, is sensitive to the world around it, grows and reproduces.
   Car: moves, needs fuel, burns fuel to make energy available (respires), produces waste gases (excretes), so it could seem to be alive BUT a car cannot reproduce itself to make more cars, or grow and get bigger. It is not sensitive to the world around it
4. You must observe the specimen over a period of time; record your observations; provide some water; look for signs of life – movement, feeding, growth, reproductive structures appearing, etc., responses to its surroundings, etc.

8

Classifying life Lesson 2

1.2 Investigating living organisms: yeast

TWS

Student Book pages 22–23

kerboodle
Online resources
- Worksheet 1.2.1 Practical investigation of respiration in yeast
- Worksheet 1.2.2 Which apparatus is best?
- Interactive test

Prior learning
- Collect evidence and data to test ideas including predictions
- Choose which equipment to use

Extension
Students produce a worksheet for another class to use, giving the method for an investigation into the effect of temperature on the rate of respiration in yeast. The worksheet must be for one of the successful methods used by students described on page 23. Remind students that their method must be written clearly so that other students can follow it and complete the practical safely.

Homework
Workbook page 7

Key words
evaluate, controlled variables

Objectives
- Plan an investigation, thinking about the variables
- Realise that not all investigations can be fair tests

Overview
This lesson helps students to develop their skills in Thinking and working scientifically. It provides an account of how different students approach an investigation into the effect of temperature on respiration in yeast. A worksheet is provided for students to carry out a version of this investigation practically. Students could do the practical instead of this spread, do the practical and compare their methods with those in this spread, or adapt the practical to follow one of the methods given in the spread. If practical work is not possible, then students simply complete this spread.

Activities
- **EITHER:** Introduce the lesson as a practical investigation into the effect of temperature on how quickly yeast respire. Question students for ideas on how the investigation might be carried out. Let them look at pages 22–23 and then give them **Worksheet 1.2.1**.
- Students carry out the practical investigation. At the end of the practical work, you can choose to put all of the class results together to give more valid and reliable data or to leave individual students/groups of students to use their individual data.
- **OR:** Work through pages 22–23 with students. Show students dried yeast and mix it with a little sugar and warm water. During the lesson, the yeast will begin to respire and the mixture will become frothy OR bring in a sourdough starter. Ask students to predict how temperature will affect the rate at which the yeast respire.
- Give students **Worksheet 1.2.2**. They read through the section headed **Choosing the equipment** and fill in the worksheet, deciding the advantages and disadvantages of the three sets of apparatus shown. Encourage students to share ideas at the end.
- Read through the section titled 'Controlling the variables' with students. Ask what is meant by a variable (something that changes) and ask students the different between a controlled variable (things that are kept the same during the investigation) and the independent variable (the variable you change in the investigation). Have a class discussion about which student is likely to obtain the most accurate, precise and reliable results (make sure that students know what each of these terms mean. If necessary, refer students to pXX Planning and carrying out an investigation.)
- Students then read to the end of the content and answer the questions.

1.2 Student Book answers

1. Any two variables from: the amount of yeast, the amount of sugar or the amount of water in the flask.
2. Evidence A = Ali, evidence B = Khalil, and evidence C = Hana.
3. 40 °C
4. The amount of yeast, the temperature.
   Students could choose any one of the three that worked best – justify why e.g. use the gas syringe as it gave the most precise measurements. Students must state that they should use the optimum temperature for the yeast, 40 °C, and that the same temperature must be used for all the different nutrients tried.

9

There is one lesson plan for every unit in the Student Book, including Thinking and working scientifically, Science in context, and extension pages. Each lesson plan suggests activities for use in the classroom linked to the topics covered on the Student Book spread.

Each lesson plan begins with a reference to the pages of the Student Book that it covers and a summary of their objectives.

The *Overview* section of the lesson plan reviews what the suggested activities will cover to fulfil the learning objectives. Here you will also find advice and tips about common misconceptions, what you may need to review from the Cambridge Primary curriculum framework or previous lessons, and suggested questions for a class discussion.

The *Activities* section of the lesson plan lists several different activities that can be used in the classroom. These activities include fun and engaging demonstrations, interesting practical ideas, group work suggestions, reading and research activities, and ways to explore a novel topic using models, games, class discussions or Internet research.

Lesson plans that are matched to Thinking and working scientifically and Science in context units include activities that encourage students to use the skills they are learning about by planning and carrying out their own investigations, analysing data, and drawing conclusions individually or as part of a group.

Most of the lessons have suggested *Extension* activities to stretch your strongest students and help prepare them for the step up to Cambridge IGCSE®. Some of these could be carried out in class, whilst others could be set as homework.

Every content unit in the Student Book is matched to a page in the Workbook. At the end of each lesson plan the corresponding workbook page is suggested as *Homework*.

There are lots of extra resources on Kerboodle to accompany every lesson plan. These are listed in the *Kerboodle online resources* box at the top of the page, and suggestions on how they can be used are given in the *Activities* section of the lesson plan.

Finishing each unit are the answers for all of the questions in the Student Book for quick reference in the classroom.

Using Kerboodle

The Kerboodle that accompanies this book can be subscribed to separately. The course contains formative and summative assessment resources to support you in the classroom and which can be assigned for homework.

Every lesson plan in the book has a supporting worksheet available on Kerboodle as a PDF, for easy printing, so that you can tailor the lesson to your classroom and your students' needs.

Every unit in the Student Book is supported by an end-of-lesson interactive formative test. These tests are assignable to individual students and their results can be logged in the digital markbook.

Glossary quizzes for each chapter focus on the scientific vocabulary introduced in that chapter. These will help support lower ability students or students that speak English as an additional language.

The Kerboodle course also includes a wealth of summative assessments for end of chapter knowledge consolidation, end of year tests, and recreating full Checkpoint papers. We have worked with Cambridge examiners and subject experts to help reflect the new Checkpoint programme of study.

Contents

Stage 7

Stage 8

1.1 What is life?

Student Book pages 20–21

kerboodle

Online resources

- Worksheet 1.1.1 Characteristics crossword
- Worksheet 1.1.2 Extension: Living or non-living?
- Interactive test

Prior learning

- Know that there are living things and things that have never been alive
- Know that there are life processes that are common to all living things

Extension

Worksheet 1.1.2 Students consider the ways in which a car is like a living organism, and why it is not alive.

Homework

Workbook page 6

Key words

biology, biologist, characteristics, respire

Objective

- Describe the seven characteristics of living organisms

Overview

This lesson introduces your students to biology, focusing on our definition of a living organism. The pages also show students some of the variety of living organisms they will be studying. It picks up the basic ideas on the differences between 'alive' and 'never alive' that they have met earlier in the school. Students begin to develop key vocabulary to support their biology studies going forward through Stages 7–9 and on to IGCSE.

Activities

- Introduce biology as the study of living organisms. Bring three examples to the classroom: 1) a living organism, animal or plant; 2) a piece of metal which has never been living; 3) a piece of wood or a fossil which was once living. Use them to discuss the idea of living organisms, non-living (never living) things or once-living things.
- Ask students to list what they would look for to decide if something is living or not.
- Read through Unit 1.1 with students, going through each of the characteristics of living organisms in turn. Make sure that students understand that not all living things move their whole bodies around, and that some don't look very alive! See how many they identified in their initial lists.
- Living or non-living? Ask students to plan how they would investigate an unknown sample to see if it is living or non-living.
- Give students **Worksheet 1.1.1**. They should do the crossword with their text books closed – only open them if they cannot complete the crossword.
- If time, give students a variety of materials collected locally, including living materials such as mosses, lichens, plant leaves or even invertebrates, along with non-living or once-living material. They should make a table to identify each sample as living, non-living or once living.

1.1 Student Book answers

1. Movement, reproduction, sensitivity, growth, respiration, excretion and nutrition.
2. Respiration is a process which makes the energy in food available to organisms to use. It is needed for all of the other characteristics of life to take place.
3. Cat: moves, eats food, respires, excretes, is sensitive to the world around it, grows and reproduces.

   Car: moves, needs fuel, burns fuel to make energy available (respires), produces waste gases (excretes), so it could seem to be alive BUT a car cannot reproduce itself to make more cars, or grow and get bigger. It is not sensitive to the world around it
4. You must observe the specimen over a period of time; record your observations; provide some water; look for signs of life – movement, feeding, growth, reproductive structures appearing, etc., responses to its surroundings, etc.

1.2 Investigating living organisms: yeast

TWS

Student Book pages 22–23

kerboodle

Online resources

- Worksheet 1.2.1 Practical investigation of respiration in yeast
- Worksheet 1.2.2 Which apparatus is best?
- Interactive test

Prior learning

- Collect evidence and data to test ideas including predictions
- Choose which equipment to use

Extension

Students produce a worksheet for another class to use, giving the method for an investigation into the effect of temperature on the rate of respiration in yeast. The worksheet must be for one of the successful methods used by students described on page 23. Remind students that their method must be written clearly so that other students can follow it and complete the practical safely.

Homework

Workbook page 7

Key words

evaluate, controlled variables

Objectives

- Plan an investigation, thinking about the variables
- Realise that not all investigations can be fair tests

Overview

This lesson helps students to develop their skills in Thinking and working scientifically. It provides an account of how different students approach an investigation into the effect of temperature on respiration in yeast. A worksheet is provided for students to carry out a version of this investigation practically. Students could do the practical instead of this spread, do the practical and compare their methods with those in this spread, or adapt the practical to follow one of the methods given in the spread. If practical work is not possible, then students simply complete this spread.

Activities

- **EITHER:** Introduce the lesson as a practical investigation into the effect of temperature on how quickly yeast respire. Question students for ideas on how the investigation might be carried out. Let them look at pages 22–23 and then give them **Worksheet 1.2.1**.
- Students carry out the practical investigation. At the end of the practical work, you can choose to put all of the class results together to give more valid and reliable data or to leave individual students/groups of students to use their individual data.
- **OR:** Work through pages 22–23 with students. Show students dried yeast and mix it with a little sugar and warm water. During the lesson, the yeast will begin to respire and the mixture will become frothy OR bring in a sourdough starter. Ask students to predict how temperature will affect the rate at which the yeast respire.
- Give students **Worksheet 1.2.2**. They read through the section headed **Choosing the equipment** and fill in the worksheet, deciding the advantages and disadvantages of the three sets of apparatus shown. Encourage students to share ideas at the end.
- Read through the section titled 'Controlling the variables' with students. Ask what is meant by a variable (something that changes) and ask students the different between a controlled variable (things that are kept the same during the investigation) and the independent variable (the variable you change in the investigation). Have a class discussion about which student is likely to obtain the most accurate, precise and reliable results (make sure that students know what each of these terms mean. If necessary, refer students to pXX Planning and carrying out an investigation.)
- Students then read to the end of the content and answer the questions.

1.2 Student Book answers

1. Any two variables from: the amount of yeast, the amount of sugar or the amount of water in the flask.
2. Evidence A = Ali, evidence B = Khalil, and evidence C = Hana.
3. 40 °C
4. The amount of yeast, the temperature.

   Students could choose any one of the three that worked best – justify why e.g. use the gas syringe as it gave the most precise measurements. Students must state that they should use the optimum temperature for the yeast, 40 °C, and that the same temperature must be used for all the different nutrients tried.

1.3 Classification and species

Student Book pages 24–25

kerboodle

Online resources

- Worksheet 1.3.1 An animal alphabet
- Worksheet 1.3.2 Species or not?
- Worksheet 1.3.3 Naming cats
- Worksheet 1.3.4 Extension: The same species or different?
- Interactive test

Prior learning

- Use simple identification keys
- Know the main differences between animals and plants

Extension

Students use **Worksheet 1.3.4** to determine whether two clouded leopards belong to the same or different species. Expected answers include: **1** They have similar tail and body lengths, cloud markings and teeth; they have different coloured coats, their cloud markings are a different size and colour and they have at least 40 different genes. **2** They are different species because their genes are so different. **3** They have different species names so scientists classify them as different species. They share the same first name because they are closely related.

Homework

Use **EITHER Worksheet 1.3.1** or **Worksheet 1.3.3**, if not used in class, **OR** Workbook page 8.

Key words

classification, species, hybrids

Objectives

- Explain why we classify living organisms
- Describe a species as a group of organisms that produce fertile offspring

Overview

In this lesson, students develop an understanding of the biological principles of classification, including why it is so useful. They are introduced to the main groups of living organisms arranged in a classification tree. Students learn a biological definition of a species as the smallest working group of organisms in a basic classification system. They discover the binomial naming system for species and explore the value of this system to biologists.

Activities

- Give a brief introduction to the variety of life on Earth. **EITHER:** Ask each student to name a type of living organism and write them all down on your board **OR:** Ask each student to write down as many different types of living organism as they can **OR:** Give students **Worksheet 1.3.1**. This challenges students to list an animal for every letter of the Roman alphabet. Here is an example of each, but students will think of many more: antelope, bear, cheetah, donkey, elephant, fox, giraffe, hippopotamus, iguana, jaguar, koala , lion, meerkat, newt, orang-utan, panda, quetzel (a beautiful bird), rhinoceros, seahorse, tiger, umbrellabird, vulture, wolf, X-ray tetra fish, yak, zebra. Alternatively, this could be used as a homework exercise or as an extension if students also try to find the scientific name for each type of animal they list.
- Work through Fig 1.3.B on page 24 and discuss a basic classification of living things. Students draw their own classification trees with the names of the different groups of organisms.
- Explain the concept of species – ask students to give some examples. Make sure they understand the difference between a type of animal or plant and a specific species, and give the biological definition of a species.
- Use **Worksheet 1.3.2** to support understanding of species. Students use picture cards to match parent animals and fertile or infertile offspring. Use their findings to give a list of which animals are the same species and which are different species. [Same species: Dalmatian and Boxer dogs; Persian and Siamese cats. Different species: lion and tiger, donkey and horse, zebra and horse.]
- Read through the paragraph headed 'Naming species' with students. Ask students to suggest why scientific names are in Latin. [Answer: when it was originally used, Latin was widely used in universities; Using Latin avoids using any living languages, so no-one is disappointed that their language hasn't been chosen; everyone everywhere can use the same names.]
- **Worksheet 1.3.3** provides good support for this activity, in which students find and use the species names of animals mentioned in the spread or that are familiar to them.

1.3 Student Book answers

1. Sorting organisms into groups based on their characteristics. Important so we know how many different types of organisms there are.
2. A group of organisms that can reproduce to produce fertile offspring.
3. Mules are infertile – they cannot breed and so they are not a species.
4. So biologists all over the world know which organisms are being referred to. It avoids the confusion of using local names.
5. There are many places where people haven't been and so new species have not been discovered. Organisms can be very small, or very hard to find, or live in places where there are huge numbers of different species so they haven't all been identified and named; any other sensible answer.

1.4 Classifying Invertebrate animals

Student Book pages 26–27

kerboodle

Online resources

- Worksheet 1.4.1 Identifying invertebrates
- Worksheet 1.4.2 Practical invertebrate identification
- Worksheet 1.4.3 Identifying pond life
- Worksheet 1.4.4 Extension: An invertebrate collection
- Interactive test

Prior learning

- Sort living things into groups, using simple features

Extension

Give students **Worksheet 1.4.4**. This is an opportunity for students to work together in groups to produce a valuable resource. Each member of a group takes one of the invertebrate groups and follows the instructions on the card to find out more about it. The work of all the groups can be put together to produce a resource about the different invertebrate groups.

Homework

Workbook page 9

Key words

invertebrates, insects, spiders, arachnids, crustaceans, myriapods

Objectives

- Describe a species as a group of organisms that can reproduce to produce fertile offspring
- Sort group and classify organisms through secondary information

Overview

Students become familiar with the main groups of the invertebrates as part of their work on classification. If possible, bring in some living examples of invertebrates or find some good illustrations. This underpins later work on other areas of biology such as ecology, decomposition and pollination and is an important foundation for IGCSE Biology.

Activities

- Have a quick quiz asking students about the basics of classification from the previous lesson, reminding them of the main groups of animals and plants and the scientific definition of a species. Introduce the invertebrates – more species than vertebrates, vital for human life on Earth through roles as pollinators, etc.
- Give students a copy of **Worksheet 1.4.1** and work through pages 26–27 with students. For each invertebrate group, encourage students to think of local examples they are familiar with. Fill in each group on the worksheet as it is described and discussed.
- If living or preserved specimens of invertebrates are available, give students **Worksheet 1.4.2** and ask them to draw and identify the invertebrate specimens. Take note of the Safety and Animal welfare warnings.
- If no specimens are available, or if students have time, give them **Worksheet 1.4.3**, which provides drawings of a number of pond invertebrates and provides a key for identifying the animals.

1.4 Student Book answers

1. An animal without a backbone.
2. 70% invertebrates, 60% are arthropods.
3. Order in the table does not matter.

Classification group	Characteristics
Arthropods	Hard outer skeleton and jointed legs. Divided into smaller groups of insects, spiders, crustaceans and myriapods
Echinoderms	Spiny skins, star or ball shapes
Molluscs	Muscular bodies and no legs but may have tentacles. Often have a shell inside or outside the body
Cnidaria	No legs and stinging tentacles around their mouths
Worms	No legs and long bodies – may be flat, round or segmented

4. It is an arthropod – it has a hard outer covering – but it is not an insect. Insects have 6 legs and antennae. This animal has 10 legs and eyes on stalks which tells us it is a crustacean.

1.5 Simple keys

TWS

Student Book pages 28–29

kerboodle

Online resources

- Worksheet 1.5.1 Invertebrate cards
- Worksheet 1.5.2 Find the 7-spotted ladybird!
- Worksheet 1.5.3 Arthropod key
- Worksheet 1.5.4 Extension: A fish key table
- Interactive test

Prior learning

- Sort living things into groups, using simple features

Extension

Use **Worksheet 1.5.4** to turn the fish key from Fig 1.5 G into a table key. Students can also display the key on **Worksheet 1.5.3** as a table.

Homework

Workbook page 10

Key words

dichotomous key, ladybirds

Objectives

- Use dichotomous keys to classify species and groups of related organisms
- Sort, group and classify organisms using keys

Overview

This lesson helps students to develop their skills in Thinking and working scientifically. Students are introduced to the concept of dichotomous keys and how they help biologists identify a wide range of organisms in the field and in the laboratory. This lesson develops key skills of classifying and using keys, which students will build on throughout this chapter and in their IGCSE Biology courses. There are opportunities for students to use and to build keys in both the Student Book and in the worksheets.

Activities

- Ask students to recall the main groups of invertebrates. Then ask students to name as many examples of a particular group as they can e.g. butterflies and moths. Alternatively, show them examples of several different species of one invertebrate group. Discuss the difficulty of identifying different organisms accurately.
- Read through the sections titled What is a key? and How does it work?
- Work as a class to identify the six fish shown in Fig 1.5.A.
- Students work independently through the activity on page 27, identifying the three species of ladybirds given.
- Students then undertake an identification exercise based on a wide range of invertebrates, using **Worksheets 1.5.1**, **1.5.2** and **1.5.3**. To get the full use from **Worksheet 1.5.1**, it should be printed out in colour (if mounted on card or laminated these cards can be reused for several years). Students classify the 24 arthropod images from **Worksheet 1.5.1** using the instructions on **Worksheet 1.5.2** and using the classification key on **Worksheet 1.5.3**.
- The correct groups are
    - o **Other arthropods:** 4 crab; 14 millipede; 19 spider; 21 shrimp: 23 mite
    - o **Other insects:** 1 ant; 2 flea; 6 butterfly; 8 fly; 11 cockroach; 15 water boatman; 16 wasp; 18; dragonfly
    - o **Other beetles:** 3 water beetle; 9 stag beetle; 17 goliath beetle; 24 wood-eating beetle
    - o **Other ladybirds:** 7 unspotted ladybird; 5 striped ladybird; 20 brown ladybird
    - o **Other spotted ladybirds:** 10 water ladybird; 12 14-spotted ladybird; 22 11-spotted ladybird
    - o **7-spotted ladybirds:** 13 7-spotted ladybird.

1.5 Student Book answers

1. A way of identifying organisms using a series of branching questions.

2. 1. arthropod; insect; beetle; ladybird; spotted ladybird; *Coccinella septempunctata*

2. arthropod; insect; beetle; ladybird; spotted ladybird; *Adilia bipunctata*

3. arthropod; insect; beetle; ladybird; *Cycloneda sanguinea*

1.6 Classifying vertebrates

Student Book pages 30–31

kerboodle

Online resources
- Worksheet 1.6.1 Features of vertebrate groups
- Worksheet 1.6.2 Vertebrate groups
- Worksheet 1.6.3 Vertebrates
- Worksheet 1.6.4 Animal matching
- Interactive test

Prior learning
- Sort living things into groups, using simple features

Extension

Students investigate an animal that interests them from one or more of the animal groups and produce a poster showing how their chosen vertebrate has the characteristics of both a vertebrate and of the group it belongs to – fish, amphibian, reptile, bird or mammal.

Homework

If not completed in class, students do **Worksheet 1.6.4** or Workbook page 11.

Key word

vertebrates

Objectives

- Describe a species as a group of organisms that can reproduce to produce fertile offspring
- Sort, group and classify organisms through secondary information

Overview

In this lesson students become familiar with the characteristics of the vertebrates – animals with backbones. It is important to remind them of the differences between the invertebrates and the vertebrates, and to focus on both the similarities and the differences between the different vertebrate groups. Students should recognise that humans are both vertebrates and mammals.

Activities

- Ask students the difference between invertebrates and vertebrates. Reinforce a correct answer. Then ask if students know the different groups of vertebrates. Gauge current knowledge and understanding from the replies. It is an excellent learning tool to have at least one vertebrate skeleton available to show students the key vertebrate features. Emphasise the importance of the backbone to protect the spinal cord.
- Students should read through pages 31–32 and complete **Worksheet 1.6.1** as they read through each group.
- Students then complete **Worksheet 1.6.2** without using their Student Books as far as possible to reinforce their learning.
- Reinforce knowledge of the vertebrate groups using the activity based around **Worksheet 1.6.3**. Print these and give each student at least one animal
- Have five tables/desks each labelled with one of the vertebrate groups. Students place their animal in the group where they think it belongs. Check that all the animals are in the correct 12 groups (each group should have animals labelled A–F in it). Work through the main features of the vertebrate groups, asking students to raise a hand when you mention a feature relevant to their particular animal. Finally, ask students to summarise the features of each vertebrate group.
- Students complete **Worksheet 1.6.4** matching all the main animal groups to their characteristics to reinforce knowledge and understanding on animal classification.

1.6 Student Book answers

1. **a.** reptile **b.** fish **c.** reptile **d.** mammal

2. bat, a mammal but front limbs are wings and it can fly
whale, a mammal but lives in water and has very little hair
snake, a reptile but has no visible legs – slithers on the ground
penguin bird, has wings and feathers but cannot fly – swims instead.

3. Any sensible choices but must justify e.g. any example of a fish or mammal, highlighting the features of the group and classifying the chosen example in the correct group. Any fish e.g. tilapia – lives and breeds in water, has scales, gills, body temperature same as surroundings = <u>fish</u>;
any whale/dolphin, etc. – lives and breeds in water, breathes air using lungs, has small amount of hair, gives birth to live young and feeds them milk, controls its own body temperature = <u>mammal</u>.

4. Any example of a reptile, bird or mammal highlighting the features of the group and classifying the chosen example in the correct group.

1.7 Classifying plants

Student Book pages 32–33

kerboodle

Online resources

- Worksheet 1.7.1 Plant classification
- Worksheet 1.7.2 Plant classification crossword
- Worksheet 1.7.3 Plant cards
- Worksheet 1.7.4 Extension: Why aren't algae plants?
- Interactive test

Prior learning

- Sort living things into groups, using simple features
- Recognise the main features of flowering plants

Objectives

- Describe a species as a group of organisms that can reproduce to produce fertile offspring
- Sort, group and classify organisms through secondary information

Overview

This lesson reinforces all of the general principles of classification that are important for students to grasp and demonstrate right through to IGCSE. It looks at the classification of plants. There is often a surprising level of ignorance about plants and their importance among students, so this is a key lesson for you to begin to address this problem. For example, many students may be completely unaware of the non-flowering plants.

Activities

- Show a selection of objects made from plants e.g. cotton clothing, nuts, compost, rope, charcoal or wood, chocolate, coffee, saffron, etc. Ask students for anything else they can think of that is made from or depends on plants.
- Show students a selection of plants or parts of plants – ideally including a piece of conifer, a fern and a moss as well as flowering plants. Ask questions such as: What sort of plant is this? Where would you find plants like these? Name three plants that produce fruits we eat. This will give you a feel for the level of knowledge of your students.
- Remind students that plants as well as animals are classified with Latin names e.g. rice is *Oryza sativa*, date palms are *Phoenix dactylifera* and figs are *Ficus carica*.
- Give students **Worksheet 1.7.1**. Read through pages 32–33, one plant group at a time. Show students examples of the different types of plants – ideally give them examples on their desks. Complete the worksheets as each group is discussed.
- Students use **Worksheet 1.7.2** to check their knowledge of plant groups.
- Give groups of students sets of plant cards made from **Worksheet 1.7.3**. Students should sort the cards into the correct groups. They could colour the cards and use them to make a large poster showing the features of the main non-flowering and flowering plant groups with their features highlighted.

Extension

Use **Worksheet 1.7.4** to challenge students to realise that classification changes. Algae were classified as plants for many years but now they are classified in another group of organisms, the Protista. Students are asked to find out why this change came about.

Homework

Workbook page

Key words

flowering plants, seed plants, non-flowering plants

1.7 Student Book answers

1. Different variations possible – one example would be:

Non-flowering plants		Seed plants	
Liverworts and mosses (+ drawing)	**Ferns (+ drawing)**	**Conifers (+ drawing)**	**True flowering plants (+ drawing)**
Small	Can grow big	Can grow very large	Can grow very large
No roots stems or leaves	Have roots, stems and fronds	Roots stems and needle-like leaves	Roots, stems and leaves
No water carrying vessels	Some water carrying vessels	Water carrying vessels	Water carrying vessels
No flowers	No flowers	Simple flowers	Flowers pollinated by animals or wind
Reproduce using spores	Reproduce using spores	Reproduce using seeds carried on cones	Reproduce using seeds found in fruits

2. **Plant A** = Fern　　**Plant B** = Conifer 15
Plant C = Liverwort (or moss)　　**Plant D** = True flowering plant

3. Grass, fern, flowering tree = Fern – the others are flowering plants.
moss, grass, fern = Grass – the others are non-flowering plants. liverwort, moss, fern = Fern – the others do not have any stems, roots or leaves.
pine tree, date palm, rice = Pine tree – it is a conifer, the others are true flowering plants.

1.8 Making your own identification keys

TWS

Student Book pages 34–35

kerboodle

Online resources

- Worksheet 1.8.1 Making an aquatic vertebrates key
- Worksheet 1.8.2 Making a reptile and amphibians key
- Worksheet 1.8.3 Making a plant key
- Interactive test

Objective

- Sort, group and classify organisms using keys

Overview

This is a lesson that helps students develop their skills in Thinking and Working Scientifically.

Work with students on the skills needed to make and use an effective key. Students evaluate the scientific methods of the students described on pages 34–35. The worksheets provide students with the opportunity to make their own keys and to try them out on their fellow students in class.

Activities

- Explain to students that they will be thinking about the process of making a key which enables other people to identify individual organisms. This is about the way scientists work, not specifically about the different plant and animal groups.
- Ask students to look carefully at the plants in Fig 1.8 A and identify them as much as possible from the photographs. Brainstorm which features different students would highlight to help others identify particular plants and produce a class list of observations for each plant.

Prior learning

- Sort living things into groups, using simple features
- Recognise the main features of flowering plants

Extension

Students take the keys produced on page 35 with the keys they have made themselves during the lesson and convert them all to tables of questions as shown in Unit 1.5.

Homework

Workbook page 13

- Compare points raised by the class with the lists produced by Laila and Omar. Encourage students to look for strengths and weaknesses in all three lists.
- Work through the keys on page 35 with students to see how effective they are.
- Students are encouraged to draw a classification key for the plants shown on page 34, and to produce a key for their own choice of organisms.
- Alternatively give students **EITHER Worksheet 1.8.1 OR Worksheet 1.8.2 OR Worksheet 1.8.3**. Each has illustrations of six named organisms (one fish, one reptile and amphibian, one plant) and their Latin names on the second page. Students use these to make a classification key. Encourage students to exchange keys and try out each others' to see which are easiest and most effective to use. The use of Latin species names means that students cannot guess which is which and need to use the keys properly.

1.8 Student Book answers

1. A. Non-flowering plant: fern
   B. Seed plant; flowering plant: oak tree
   C. Non-flowering plant: moss
   D. Seed plant; conifer: cypress tree
   E. Seed plant; flowering plant: torch ginger
   F. Seed plant; flowering plant: golden coconut palm

2. He forgot that flowering plants don't always have flowers on them/he forgot that trees are seed plants/any other sensible points

3. e.g. below – others are possible

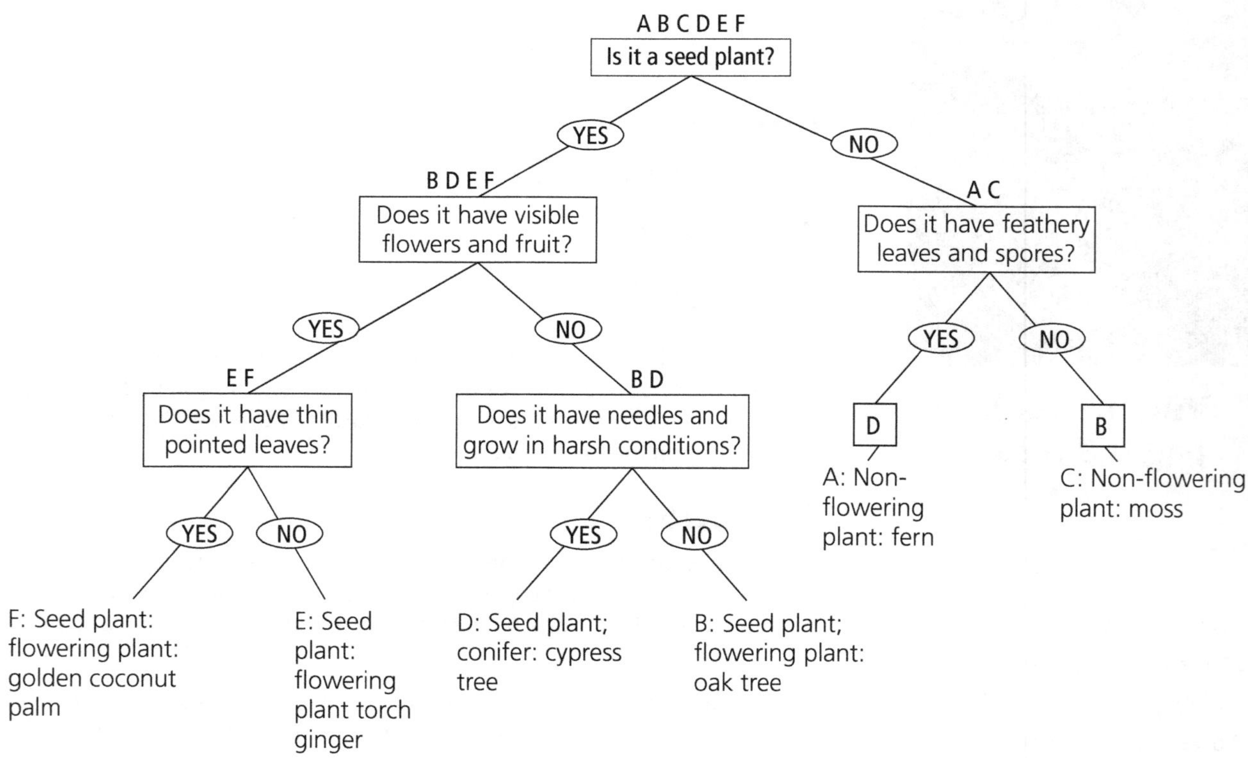

4. Look for students choosing an interesting range of organisms, identifying them correctly themselves, then working out a good key with sensible questions.

1.9 Are viruses living?

SIC

Student Book pages 36–37

kerboodle

Online resource

- Worksheet 1.9.1 Viruses: living or non-living?
- Interactive test

Prior learning

- Decide what to do to try answer a science question
- Test an idea based on scientific knowledge

Extension

Students investigate parasites and write a paragraph explaining how viruses act as parasites in the cells they invade.

Homework

Workbook page 14

Key words

genetic material, parasites

Objectives

- Discuss reasons for classifying viruses as living or non-living
- Evaluate issues which require scientific understanding

Overview

This lesson helps to students develop their understanding of Science in context.

It introduces students to the important idea that science is not about facts – it is about ideas which develop through discussion and evidence. To achieve highly at IGCSE, students need to understand the nature of science and that scientists do not always agree. This is a good starting point for developing that understanding.

Activities

- Ask students to give you as many facts as possible about viruses. This enables you to see how much, or little, students know about this topic.
- Read page 36 with students and point out the very geometric shapes of viruses. Explain that they are much smaller than bacteria, which students may have heard of, and that they all cause disease.
- Ask students to list the seven characteristics of living things and write them on the board for everyone to see.
- Explain how science moves forward as a result of debate and questioning among scientists before students read the spread in the Student Book and use **Worksheet 1.9.1**.
- Ask students to vote on the question: Are viruses living organisms? Record the numbers voting 'Yes' and 'No'.
- Give students time to prepare and then chose several students to give their points of view. Make sure that the same number of students speak for both ideas.
- Take another vote on the same question – how many students, if any, have changed their minds?
- In any time remaining, students work through the end of spread questions.

1.9 Student Book answers

1. Tiny particles which invade the bodies of living organisms and turn them into virus-making factories.
2. Any two e.g. colds, flu, COVID-19, measles.
3. Act as parasites on other living organisms. Reproduce using all the resources of the host organism. Found everywhere there are other living organisms.
4. Do not carry out the characteristic processes of life apart from reproduction which they can only do inside a living organism. Can be stored for many years and still remain infective.

Extension 1.10 Moving classification forward

Student Book pages 38–39

kerboodle

Online resources

- Worksheet 1.10.1 Butterfly observations
- Worksheet 1.10.2 Big cat keys
- Worksheet 1.10.3 Extension: Instructions for biologists
- Interactive test

Prior learning

- Know that scientists have combined evidence with creative thinking to suggest new ideas and explanations

Extension

Students use **Worksheet 1.10.3** to plan: a) how they would investigate a sample brought back from a space mission to see if it is alive; and b) how they would classify an organism brought out from a Central American rain forest. They should use all of the skills they have learned so far.

Homework

Workbook page 15

Key words

biodiversity, genetic material

Objectives

- Describe a species as a group of organisms that can reproduce to produce fertile offspring
- Describe how people develop and use scientific understanding

Overview

In this extension lesson, students are moving towards their GCSE studies by adding to the knowledge of classification and keys that they need for CLSS. They are beginning to look at how scientists use genetic information to support classification. They are also developing a more holistic view of the biology they are learning, as they discover the value of classification in helping us measure biodiversity and so measure changes in ecosystems and populations.

Activities

- Begin by asking students to reflect on what they have learned about classification. Careful questioning will show you if they understand the importance of observation in classifying organisms. Ask why they think classification has always been based on observation. [This checks that they understand that this was basically all we had until relatively recently.]
- Give students **Worksheet 1.10.1** and ask them to complete the dichotomous (branching) key based on detailed observations of butterflies.
- Students then read down to the end of the section on clouded leopards. They use **Worksheet 1.10.2** to develop understanding of how a combination of observational and genetic evidence enables scientists to build the best possible key. Students start by using limited observational information to build a functional key. As they work through the worksheet, they are given more observational and genetic information which allows them to develop a much more effective key, based on the biological relationships between the big cats.
- Read through the rest of the information with students and spend time examining the data. Students answer question 4 at the end of the page about the fall in the number of vertebrate species over time.
- Students produce an illustrated poster or leaflet explaining the links between classification and what we know about biodiversity.

1.10 Student Book answers

1. The variety of types of organisms
2. Any 2 from: observation/analysing the genetic material/breeding them to see if they produce fertile offspring.
3. It allows us to measure changes in biodiversity – we need to know what species we have so we can observe if they are increasing or dying out.
4. **a.** The trend in the graph is downwards – the number of vertebrate species kept falling steadily.
   b. 40%
   c. 60%

1.11 Review answers

Student Book pages 40–41

Student Book answers

1. There are seven characteristics of living things and scientists should look for evidence of some or all of them taking place: movement, respiration, sensitivity, growth, reproduction, excretion and nutrition. **[3]**
2. **a.** They show that the dough containing yeast increased in height from 12 mm to 42 mm over two hours. The dough without yeast stayed the same height. **[2]**
   b. Yeast is a living organism. It respires and produces carbon dioxide which makes the dough rise. **[2]**
   c. Any one from: Put the dough in a warmer place/put more sugar in the dough/put more yeast in the dough. **[1]**
3. **a.** So scientists all over the world know which organisms they are talking about. **[1]**
   b. Because the bears are similar so they are in the bear group together. **[1]**
   c. Because the bears are different species. **[1]**
   d. If different species breed they form a hybrid and hybrids are usually infertile. **[2]**
4. **a.** B and C (octopus and beetle). **[2]**
   b. They don't have backbones. **[1]**
   c. A and D (snake and tortoise). **[2]**
   d. Any two from: Dry, scaly skin/ lay eggs (reproduce) on land/ lay eggs with leathery shells/body temperature varies with their surroundings/ breathe air using lungs. **[2]**
5. **a.** All have a backbone/are vertebrates. **[1]**
   b. A = mammal, B = mammal, C = bird, D = amphibian. **[4]**
6. Give birth to live young, feed young on milk made in the body, have fur. **[3]**
7. **a.** Invertebrates/arthropods **[2]**
   b. A and C **[1]**
   c. Insects – they both have 6 legs (or any other insect feature visible in drawings). **[2]**
   d. Arachnid – it has 8 legs. **[2]**
8. **a.** Moss **b.** Fern **c.** Conifer **[1, 1, 1]**
9. **a.** A group of organisms that can breed to produce fertile offspring. **[1]**
   b. Any two from: make detailed observations of their appearance/analyse their genetic material and compare it with other similar monkeys/see if they can breed and produce fertile offspring. **[2]**
10. Biologists have identified seven characteristics of life – movement, reproduction, sensitivity, growth, respiration, excretion and nutrition. Living organisms carry out most or all of these processes. Biologists who think viruses are living argue that they carry out these processes by being perfect parasites. They invade a host and the host organism feeds, respires, excretes, grows and moves – it carries out all the processes of life for the virus to use. The virus then uses the material produced by the host to reproduce itself. Biologists who think viruses are not living organisms argue that viruses cannot carry out any of the characteristics of a living organism on their own. When they invade an animal or plant, they take over its living processes and use them to make more viruses. This is the only way they can reproduce. Some viruses can survive for years stored in an airtight container. They still cause disease when they are released. They do not fit our model of living things. **[4]**
11. An appropriate dichotomous key. **[4]**

2.1 The building blocks of life

Student Book pages 42–43

kerboodle

Online resources

- Worksheet 2.1.1 Light microscope word search
- Worksheet 2.1.2 The light microscope
- Interactive test

2.1 Student Book answers

1. The building block of all living things.
2. A piece of equipment that uses light and lenses to magnify small objects.
3. Because most cells are too small to see with out eyes – they need to be magnified for us to observe them.
4. The order of the list does not matter:

   Eyepiece lens: The lens you look down/makes things look bigger

   Objective lens: Makes things look bigger

   Stage: Where you put your microscope slide

   Clip: Holds your specimen in place on the stage

   Mirror/light source: Provides the light you need to see your specimen

   Focusing knobs: Turn these slowly to focus the microscope so you can see the specimen clearly
5. Any from: LM magnifies to ×1000, EM magnifies to ×1 million; EM larger than LM; EM much more expensive than LM.

Objectives

- Know that all organisms are made of cells
- Most microorganisms are single cells

Overview

This lesson introduces students to cells as the basic unit of organisms. This will be the first time many students will have met the idea of cells and microscopes – it is a great opportunity for developing wonder and excitement about what they are studying. Understanding cells is key to achieving well at IGCSE, so the foundations you build through this topic will be important both now and in the future. Students begin to develop key vocabulary to support their biology studies going forward through Stages 7–9 and on to IGCSE.

Activities

- Use page 42 of the Student Book to introduce the idea of cells as the building blocks of living organisms and discuss the use of models in helping us to understand science (reference Thinking and working scientifically). Bring in any models you have that can help students understand the idea. Children's building bricks in different shapes, sizes and colours are useful.
- Carry on through pages 42–43 to explain that all of the processes of life take place within cells – and then discuss ideas of size. Students may not realise that, because they cannot see things with their eyes, they still exist. Ask students to estimate sizes and look at rulers. Show them that there are 1000 mm in 1 m. Help them use their imaginations to imagine 1000 mm in a 1 mm.
- If available, ask students to examine everyday things through a simple magnifying glass to get a practical idea of magnification.
- Now give students **Worksheets 2.1.1** and **2.1.2**. Using a real microscope and/or Fig 2.1 D, demonstrate the different parts of the microscope and describe what they do. Students complete **Worksheet 2.1.1** to familiarise them with the parts of the light microscope. Every time they find and cross out the name of the part of the microscope in the word search, they add it as a label to the diagram of a light microscope on **Worksheet 2.1.2**. Students must keep their worksheets – they will add more labelling in a later lesson.
- Read through the section on the electron microscope with students – students just need to understand that these big, expensive machines give us even more magnification and clearer images of the details inside cells.
- **Fun fact!** To help students grasp the power of microscopes, tell them that a person magnified by the best light microscope would appear to be 2.5 km tall. The same person magnified by an electron microscope would be almost 1000 km tall!
- Students answer the questions at the end of the spread.

Extension

Students measure small objects with their rulers and calculate their size in mm and μm.

Homework

Workbook page 16

Key words

cells; microns; light microscope; specimen; electron microscope

2.2
The cell story

SIC

Student Book pages 44–45

kerboodle

Online resource

- Worksheet 2.2.1 A timeline of microscopes and cells
- Interactive test

Prior learning

- Consider how scientists have combined evidence from observation and measurement with creative thinking to suggest new ideas and explanations for known phenomena

2.2 Student Book answers

1. Cells are the basic unit of life.
2. He bought fabric and made lenses so he could check his materials.
3. That a plant cell has a nucleus.
4. Theodore Schwann
5. Very few people had good microscopes with clear lenses. They couldn't see the cells Hooke and van Leeuwenhoek were describing and so wouldn't accept them. By the time Schwann proposed the cell theory, there were many good microscopes and communication between scientists was much easier. Cells were widely seen and accepted in all tissues so it was easy to agree that they are the basic units of life.

Objectives

- Understand that all organisms are made of cells
- Discuss how scientific knowledge develops over time

Overview

This lesson helps students to develop their understanding of Science in context. It summarises the early development of microscopes and looks at the role that these new technologies played in arriving at the cell theory. The content is delivered through the reports of two students. This is an excellent opportunity for students to consider the history of science and the impact of technology on our scientific understanding. In this instance, microscopes enabled us to see cells. Other examples include MRI scanners allowing us to see inside the human body, PCR machines enabling us to sequence DNA, or any local examples you may have of scientific discoveries.

Activities

- Ask students to think of concrete examples which make it easier to understand something. For example, it is easier for a small child to understand water flowing from a tap than to understand how electricity works – because we can see water. Then ask them to travel back in time over 400 years and imagine how hard it would be to accept that living organisms like people are made up of cells, when no-one had ever seen a cell.
- Introduce the students to Rameshwar and Karin and explain that they are going to look at the reports that these students have produced.
- Ask students to read through the section titled 'The history of the microscope'. Find examples of fabric magnified to show the threads, to explain why Anton van Leeuwenhoek wanted to produce a microscope, and any other examples of early drawings of cells you can find online. Students answer questions 1–3 on page 45.
- Ask students to consider how much easier it would be to visualise all living things as being built up of cells having seen the evidence from these early microscopes. Students read the information under the heading 'The cell theory' and answer questions 4 and 5 on page 45.
- Give students **Worksheet 2.2.1**. This is a collaborative exercise in which individual students or groups of students work on areas of the history of cells and develop visual resources to make a timeline which can be displayed in the classroom. You can provide large sheets of paper for their work, and arrows to link each time frame to the next. Some of the elements suggested are covered on pages 44–45. Others are not and require information from other books or the internet. Select which students are given which tasks carefully. Each student or group of students should produce at least one, and no more than three, paragraphs of writing, and an image or drawing. Encourage them to identify the step forward in scientific knowledge and understanding resulting from the development they describe.

Extension

Each student produces their own timeline of microscopy with a minimum of seven dates and different instruments. Encourage them to identify the resulting step forward in scientific knowledge and understanding.

Homework

Workbook page 17

Key word

cell theory

2.3 Animal and plant cells

Student Book pages 46–47

kerboodle

Online resources

- Worksheet 2.3.1 Animal cells or plant cells?
- Worksheet 2.3.2 Animal and plant cells under the microscope
- Worksheet 2.3.3 Plantimals!
- Interactive test

Prior learning

- Know the main differences between animals and plants

Homework

Workbook page 18

Key words

estimate, nucleus, genes, cell membrane, cytoplasm, mitochondria, respiration, cell wall, vacuole, cell sap, chloroplasts

Objectives

- Understand that all organisms are made of cells
- Identify and describe the functions of some cell structures

Overview

In this spread, students are introduced to some key features of animal and plant cells. The aim is to ensure that they are familiar with both the structures and their functions, so that they can identify and describe them in diagrams, micrographs and as they look at them on microscope slides themselves. Remind students of the size of these cells – both their absolute sizes and the comparison between animal and plant cells. Also emphasise that cells are 3-D structures, not flat 2-D things like the diagrams on the page. This is revisited in lesson 2.9.

Activities

- Prepare for the lesson, by printing out the cells shown in **Worksheet 2.3.1**. Print each one on a separate sheet of paper and mount it on card so that you can hold it up to the class OR prepare a presentation with one of the cell cards on each. Mix up the animal and plant cells that you will be using in the lesson.
- Ask students to remember and list the similarities and differences between animals and plants. Then introduce the idea that, just as animals and plants have similarities and differences, so do the different types of cells, which are the units of the organisms.
- Start by looking at the animal cell on page 46. Explain that this is a simple animal cell – similar to ones lining their mouths. Discuss with students the different structures of the cell and what they do. Ask them to indicate, by a show of hands, how many of these structures will also be found in plant cells.
- Look at page 47 and read through the information on plant cells with students. Make sure they appreciate the size of the cell, and the features that are similar to and different from animal cells.
- Carry out the activity on **Worksheet 2.3.1** with the whole class. Show them the cells and ask them to write down whether each image is of animal or plant cells. They should explain their choice in each case. Then go through the cells, telling them whether they are plant or animal, pointing out features such as regular and irregular shapes, presence or absence of cell walls, etc. Students should mark their own work (or exchange with a partner) – discover how they got on by a show of hands.
- **EITHER** Give students **Worksheet 2.3.2** and ask them to complete the activity **OR** ask students to complete Questions 1 and 2 at the end of page 47.
- Once students have completed this activity, ask them to read the final section of page 47 and then complete questions 3 and 4.

Extension

Students use **Worksheet 2.3.3** to determine whether *Euglena* and *Chlamydomonas* are animal or plant cells. Either answer is acceptable as long as it is justified by evidence. Students are then asked to identify features that tell them that these are living organisms. Only TWO features are visible so you are looking for two answers: flagella move the organisms – movement is a characteristic of living things; they have chloroplasts to make their own food – living organisms need food to supply the cells with energy.

No other answers – these are the two they can see. Emphasise the importance of reading and answering ONLY the question asked.

2.3 Student Book answers

1.

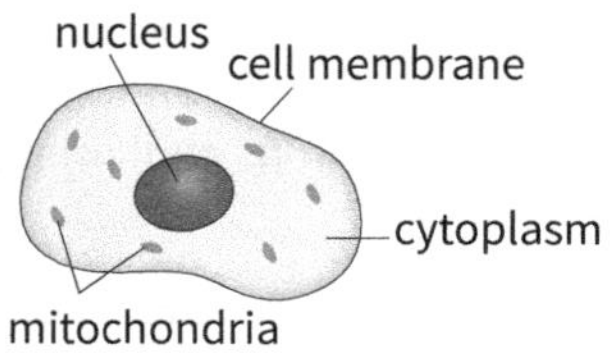

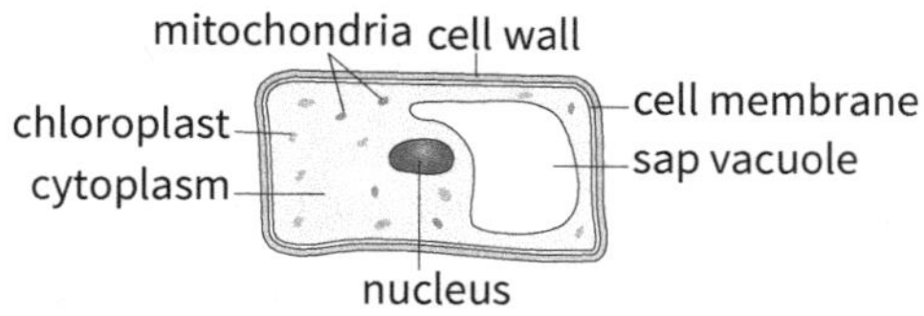

2. Order in table does not matter

Part of cell	Function	Animal, plant or both?
nucleus	Contains a set of plans to make new cells and controls everything going on in the cell	Both
cell membrane	Controls movement of substances into and out of the cell	Both
cytoplasm	Jelly-like substance where many of the chemical reactions of the cell take place	Both
mitochondria	Provide energy for the chemical reactions in the cell	Both
vacuole	Filled with cell sap, pushes against the cell wall to keep the cell firm and support the organism	Plants
cell wall	Tough outer layer which gives the cell strength. Vacuole pushes against it to keep cell firm and support organism	Plants
chloroplasts	Capture light energy for cell to make food	Some plant cells

3. Contains instructions to make a new cell; controls the chemical reactions going on in the cell
4. The cells of leaves and stem contain chloroplasts which capture the light energy needed for the plant to make food. Parts of the plant which don't get any light like the roots which grow underground don't have any chloroplasts.

2.4 Using a microscope

TWS

Student Book pages 48–49

Objectives

- Sort organisms through observation
- Collect and record observations in an appropriate form

Overview

This lesson helps students develop their skills in Thinking and working scientifically.

Used in conjunction with **Worksheets 2.4.1–3**, pages 48 and 49 enable you to introduce your students to the practical use of the light microscope. They explain how to use a simple light microscope, how to calculate magnification and – on the three worksheets – give clear instructions both for practising using a microscope and for looking at animal and plant cells, either using slides prepared by the students or using ready-prepared and stained slides. You could use a combination of the Student Book and the worksheets to make two sessions which would leave your students with a good understanding of microscope techniques, or you could deliver a single very full lesson – preferably using pre-prepared slides for time management. Alternatively, you can use the Student Book alone without practical work if microscopes are not available.

kerboodle

Online resources

- Worksheet 2.4.1 Using a microscope
- Worksheet 2.4.2 Observing animal cells
- Worksheet 2.4.3 Observing plant cells
- Interactive test

Prior learning

- Sort living things into groups, using simple features

Extension

Students complete questions 1–4 on page 49. They then investigate the use of different stains to show up different structures within animal and plant cells.

Homework

Workbook page 19

Key word

stains

Activities

- Start with a light microscope in front of you and ask students to name the parts from their earlier lessons. Now read through Amira's instructions for using a microscope. Students return to **Worksheet 2.1.2** – their labelled diagram of a microscope – and, using page 48, add annotations explaining what each part of the light microscope is used for.
- **If microscopes are available for your students:** Give students a copy of **Worksheet 2.4.1** and let them all practise using a light microscope. Give them time to get the microscopes set up to see how graph paper appears when it is magnified.
- Talk to students about drawing their observations – using pencil, and always showing the magnification. Work through the section on magnification on pages 48–49 and give students examples to try.
- Read the final section with students that explains the importance of stains. **If you do not have microscopes available:** show students a presentation of many different types of cells under the light microscope using different stains. **If no microscopes are available:** students should complete Q5, using the micrographs on page 49 as if they were seeing them down the microscope and making observational pencil drawings from them.
- Students then complete questions 1–4 in any remaining lesson time.
- **If light microscopes are available: EITHER** give students prepared slides of human cheek cells and onion cells and ask them to follow the instructions of **Worksheet 2.4.2** and **Worksheet 2.4.3** to observe and draw the cells using their light microscopes **OR** continue this work for another lesson, giving students worksheets **Worksheet 2.4.2** and **2.4.3** and providing them with the opportunity to make slides of their own cheek cells and of onion cells, which they then observe and draw. Make sure students understand how to calculate and record the magnification of the cells they observe and draw.

2.4 Student Book answers

1. Any suitable variation on: put stage at lowest position → put specimen on stage, secured by clips → put lowest power objective lens in place → look through eyepiece and turn coarse focus knob gently until the image comes into focus → adjust sharpness of image using fine focus knob → make a pencil drawing of a sample of cells. Include magnification.
2. total magnification = eyepiece lens magnification × objective lens magnification = 10 × 40 = ×400
3. Cells are often colourless. Adding stain makes it easier to see features of cells, e.g., nucleus, cytoplasm, cell walls.
4. Mitochondria are so small they cannot be seen at this level of magnification.

2.5 Specialised animal cells

Student Book pages 50–51

kerboodle

Online resources

- Worksheet 2.5.1 Similarities and differences
- Worksheet 2.5.2 Specialised animal cells
- Worksheet 2.5.3 Extension: Specialised sense cells – energy detectors
- Interactive test

Prior learning

- Sort living things into groups, using simple features

Extension

Worksheet 2.5.3 presents students with two very specialised cells and challenges them to relate the structures they can see to the functions of the cells.

Homework

Workbook page 20

Key words

multicellular, specialised cells, function, plasma, red blood cells (rbcs), haemoglobin, biconcave, neurones (nerve cells), insulation, ciliated cells, cilia

Objectives

- Explain the need for specialised cells
- Explain how the structures of specialised cells are related to their functions in the cell

Overview

Now that your students have grasped the cell theory that underpins biology and know the basic structures found in many cells, you are going to introduce the concept of specialised cells. There are several examples in the Student Book – and the more examples you find and explore with them, the better they will become at understanding the main principles. This is the ideal place to emphasise the importance of the relationship between structure and function, a key concept for future success at IGCSE.

Activities

- Use a quick quiz to recap the basic structures of animal and plant cells.
- Read the first part of page 50 with your students and discuss the need for specialised cells in multicellular organisms. Ask students to name any specialised cells they have heard of. The response depends on your class. If students have lots of ideas, encourage and praise them. If they do not know any specialised cells, help them to work out that we need cells to move our bones, carry oxygen, etc.
- Students answer questions 1 and 2 on page 51.
- Read through the content on red blood cells. Explain the unique structure of red blood cells and how this relates directly to their function in the body. Students complete Q 3 from page 51 to reinforce their learning on this point.
- Give students **Worksheet 2.5.1**. Here students are directed to consider how the structure of a neurone relates to its function. They use the content of the Student Book to help them to complete the worksheet. Check that students have completed the worksheet successfully by going through the answers with the whole class.
- Students then read the rest of page 51 to help them to complete **Worksheet 2.5.2**. If possible, prepare a short presentation to show students micrographs of several different types of specialised cells, e.g., striated muscle cells, fat cells, secretory cells, and discuss how their structures are related to their functions.
- If there is time, students can **EITHER** carry out some research to find more examples of specialised cells, relate their structure to their function and add them to their table on **Worksheet 2.5.2**, **OR** complete Q4 to reinforce their learning on neurones and ciliated cells.

2.5 Student Book answers

1. An organism made up of more than one cell, usually millions or billions of cells

2. A cell specialised to carry out a particular function in an organism

3. a. To carry oxygen around the body

b.

Feature	Function
Filled with haemoglobin	Carries oxygen
Small and flexible	Pass easily through small blood vessels to carry oxygen to the cells
No nucleus	Makes extra space for more haemoglobin to carry more oxygen
Biconcave	Gives them a big surface area pick up lots of oxygen

4. Neurones: version of Fig 2.5 C with labels:

- long structure to carry messages around the body
- branches to connect to other neurones
- insulation so electrical messages travel faster.

Ciliated cells: version of Fig 2.5 D with labels:

- cilia which beat to cause movement
- many mitochondria to supply the energy needed to make the cilia move.

2.6 Specialised plant cells

Student Book pages 52–53

Objectives

- Explain how the structures of some specialised cells are related to their functions

Overview

In this lesson, your students apply their understanding of the need for specialised cells in multicellular organisms, and reinforce the concept of the relationship between structure and function in specialised cells. You will introduce students to examples of specialised plant cells before discussing how to identify specialised features in unknown cells.

Activities

- Review the work from the previous lesson on specialised cells, asking students questions such as 'What is a specialised cell?', 'Give me an example of a specialised animal cell', before looking at pages 52–53 and discussing plants as multicellular organisms that also need specialised cells.
- Ask students to list some of the things that plants do, and then use these ideas to develop science-based ideas about the types of specialised cells plants might contain. Then read 'Cells as food factories' and 'Root hair cells' with students before they answer questions 1, 2 and 3.
- Students read through the section headed 'How to identify a specialised cell' and answer Q4 on page 53. Make sure students are aware that the principles discussed here apply to both plant and animal cells.
- This leads into the activity based on **Worksheet 2.6.1**. Ideally students work in pairs for this activity, although it can be completed individually. Each student in the pair has one of the worksheets and cuts up the six cards. Students take it in turn to show a card – their partner must match it to the correct one in their selection. The activity encourages students to integrate the appearance of a cell with its structure and function.

2.6 Student Book answers

1. Because they contain more than one cell
2. Cells found at the top of a leaf. Brick shaped so they pack in tightly. Many chloroplasts to trap as much light as possible.
3. Chloroplasts – because root hair cells grow in the soil where there is no light so they do not need chloroplasts which trap light for photosynthesis.
4. Cell A = animal cell – no cell wall or vacuole or chloroplasts. Not very active – very few mitochondria. Fat store as huge amount of fat in the middle. Cell B = plant cell – cell wall, vacuole and chloroplasts. Not a very regular shape and not very many chloroplasts so not in top layers of a leaf.

kerboodle

Online resource

- Worksheet 2.6.1 Identifying specialised cells – matching structure and function
- Interactive test

Prior learning

- Sort living things into groups, using simple features

Extension

Students develop a small poster describing one example of a specialised animal cell and one of a specialised plant cell. Extra credit is given for researching and using different examples to those given in the Student Book.

Homework

Workbook page 21

Key words

photosynthesis, palisade cells, chlorophyll, mineral salts, root hair cells

2.7 Modelling cells

TWS

Student Book pages 54–55

kerboodle

Online resources

- Worksheet 2.7.1 Making a model cell
- Worksheet 2.7.2 Extension: Using models: advice for teachers
- Interactive test

Objectives

- Identify cell structures including the cell membrane, cytoplasm, nucleus, mitochondria, cell wall, chloroplasts and sap vacuole
- Describe the similarities and differences between animal and plant cells
- Describe the strengths and limitations of a model

Overview

This lesson helps students to develop their skills in Thinking and working scientifically. It helps your students to discover the benefits and limitations of models in science in a very practical way. The concept of models is important through to IGCSE Biology and beyond, so this is a valuable foundation. This lesson gives students the opportunity to evaluate different model cells, to understand both how useful models are at helping us visualise things that we cannot normally see, and how difficult it is to create a simple model that does not reinforce misconceptions. Giving students the opportunity to create their own model of a cell, whether in class or as a homework project, reinforces that understanding.

Activities

- Begin your lesson by quickly revisiting with students the main features of animal and plant cells. Remind students of
    1. the difficulty of seeing cells even with a microscope
    2. the difficulty in visualising cells as three dimensional, not flat things like our diagrams
    3. the problems of visualising cells as jelly-filled with structures such as the nucleus and mitochondria suspended in the cytoplasm.
- Explain the idea of scientific models, as a physical way of representing things we cannot easily see – often because they are too small or too big. Explain how easily misconceptions are introduced by models – and the difficulty of overcoming those misconceptions in students. Use an example from childhood – for example, small children may be afraid of draughts because they do not know what they are, so imagine them to be monsters.
- There are many ways of using this lesson. Here are three suggestions:
- **EITHER:** Work through pages 54–55 with your students, discussing the different models made by the students. Encourage students to look for strengths in each model and possible misconceptions they might cause by answering questions 1–4 at the end of page 55.
- Have a class vote for which is the best model, then give your students **Worksheet 2.7.1**. Students design their own model cell and, if there is time, make the model as a project in class or at home.
- **OR:** Divide the class into groups or pairs. Allocate one of the three models to different groups of students and give them 10 minutes to evaluate it. Encourage students to look for strengths in each model, and possible misconceptions they could cause, using the relevant question (1, 2 or 3) from page 55 to structure their discussions.
- Students from the different groups state advantages or disadvantages of the model they have evaluated. Students vote for the cell model they think works best.
- Give your students **Worksheet 2.7.1**. Students design their own model cell and, if there is time, make the model as a project in class or at home.

- **OR:** Spend a short time summarising the models on pages 54–55, commenting on the strengths of each and pointing out potential misconceptions they might introduce and reinforce in students. Base your comments on questions 1–3 on page 55. Students vote for the cell model they think works best.
- Give your students **Worksheet 2.7.1**. Students design their own model cell and, if there is time, make the model as a project in class or at home.

Extension

Give students **Worksheet 2.7.2**. Students write a brief report to be used for training teachers. They should explain the value of using models to help students visualise and understand cells. They should also explain some of the problems with using model animal cell and plant cells and the misconceptions they can cause.

Homework

EITHER use **Worksheet 2.7.1** and complete the model cell planned and started at school, plan a model cell or make a model cell.

OR Workbook page 22.

Key word

misconception

2.7 Student Book answers

1. a. It is simple; is has an irregular shape; it shows the nucleus of the cell and the cell membrane; any other sensible point.
   b. She hasn't labelled all the structures; it has an incorrect label – animal cells do not have vacuoles; it has got green structures inside it. If these are mitochondria, they should never be coloured green because that is the colour of chloroplasts. If they are chloroplasts, they are wrong because animal cells don't have chloroplasts; it doesn't give any idea of size.
2. a. Rafi's model is really good for showing how the plant cell wall is kept strong and rigid by the inside of the cell pressing against it. It shows the difference between the cell wall – the box – and the cell membrane – the balloon. It has models of the nucleus and the chloroplasts.
   b. It has no labels to explain what anything is except for the cell wall; it has air in the balloon instead of the jelly of the cytoplasm; the balloon does not fill the shape of the box; there are no mitochondria; there is no sense of scale; any other sensible point.
3. a. It gives an understanding of the size difference between animal and plant cells; shows the difference in shape between animal cells (irregular) and plant cells (regular); shows that plant cells have a strong rigid cell wall and animal cells do not; highlights that animal and plant cells have lots of features in common as nucleus, cell membrane and mitochondria are made of the same things and are labelled in both models; any other sensible point.
   b. The cytoplasm is flat and made of modelling clay – real cytoplasm is jelly-like and structures such as the nucleus and mitochondria are floating in it. These models make them look as if they sit at the bottom of the cell. Any other sensible points.
4. Students can make any choice they wish as long as it is clearly justified – for example, it can be useful to look at a poor model like Meryl's and correct the labels and suggest ways it might be improved.

2.8 Tissues and organs in animals

Student Book pages 56–57

kerboodle

Online resources

- Worksheet 2.8.1 Animal tissues and their functions
- Worksheet 2.8.2 The main human organs
- Worksheet 2.8.3 Human organ crossword
- Interactive test

Prior learning

- Use scientific names for some major organs of the body
- Identify the position of major organs of the body
- Explain how the functions of the major organs are essential

Extension

Students **EITHER** chose an animal/human organ system and carry out some research into the organs and tissues that make it up and their functions **OR** develop a Human tissues, organs and systems word search.

Homework

Workbook page 23

Key words

cells, tissues, organs, organ systems

Objective

- Understand that cells can be grouped together to form tissues, organs and organ systems

Overview

In this lesson, you will introduce students to the concept that multicellular organisms are made up of specialised cells arranged as tissues which work together to carry out functions, and that these tissues are then organised within organs and then as organ systems. A solid understanding of these principles is fundamental to biology both for LCSS, for IGCSE and for Advanced courses beyond. This lesson and the next give you time to reinforce these concepts for your students in both animals and plants.

Activities

- Use quick-fire questions or a planned quiz to discover how many different types of specialised cells your students recall.
- Introduce the idea that large, multicellular organisms need organisation to survive. Use an analogy which your students will be familiar with – perhaps a hospital, a factory, a hotel or even a school. If everyone tried to do all the jobs, it would be chaos – and nothing would be done very well. Ask students to list as many jobs as possible in a hospital, for example, doctors diagnose patients, prescribe drugs and do operations; nurses take care of patients, give them medicines and deliver babies; orderlies move patients around and make beds, cooks prepare food, cleaners make sure everywhere is clean ....
- Read the first part of page 56 with your students, introducing the levels of organisation in multicellular organisms. Explain how each level is built up of the previous one and explain how the function develops. Students answer Q1.
- Read the section headed 'Animal tissues' with students who then complete **Worksheet 2.8 1**. In this activity, students focus on the different types of cells making up tissues found in the human skin.
- Read the section headed 'Animal organs' with students, giving them examples and talking through Fig 2.8 B which shows a range of human organs. Students complete **Worksheet 2.8.2** and **Worksheet 2.8.3**. **EITHER** as they complete each clue on the crossword, they also fill in the label on the diagram **OR** they label and annotate the diagram on **Worksheet 2.8.2** and then complete the crossword.
- Read section headed 'Organ systems in animals' with your students, making it clear that in an organ system many organs work together to carry out a large-scale function in the body such as digestion or transport. Students complete the questions at the end of page 57.

2.8 Student Book answers

1.

Level of organisation	Description
cell	Building block/basic unit of all organisms
tissue	Group of similar cells working together
organ	Several different tissues working together to carry out a particular function
organ system	Several different organs working together to carry out a particular function
whole organism	Combination of many different organs and organ systems working together to carry out all the functions of life

2. a. heart **b.** kidney **c.** liver

3. A tissue is made up of one type of specialised cells. An organ is made up of several tissues working together, so it would contain several different types of specialised cells.

4. a. To break down and digest the food taken in and remove undigested waste.

b. Most likely three: stomach – digests food; intestine – digest food/absorbs nutrients; liver – produces bile to help digest food.

2.9 Tissues and organs in plants

Student Book pages 58–59

kerboodle

Online resources

- Worksheet 2.9.1 The main plant organs
- Worksheet 2.9.2 Extension: Levels of organisation in multicellular organisms
- Interactive test

Prior learning

- Know that plants have roots

2.9 Student Book answers

1. **a.** organ
   b. organ system
   c. specialised cell
   d. tissue
   e. organ
2. Palisade cell = specialised cell; palisade tissue – many palisade cells working together to trap light energy and carry out photosynthesis.
3. Flowers are plant reproductive organs – they contain all the tissues needed to make new plants BUT plants do not reproduce all the year round and flowers are only produced when needed.

Objective

- Understand that cells can be grouped together to form tissues, organs and organ systems

Overview

In this lesson, you will reinforce the concept that multicellular organisms are made up of specialised cells arranged as tissues which work together to carry out functions, and that these tissues are then organised within organs and then as organ systems. Students apply what they learned in the last lesson on animal systems to tissues, organs and organ systems in plants. In this way, students will build the fundamental knowledge of plant biology that they need to succeed at the highest level in this course and in IGCSE Biology.

Activities

- Begin with a quick quiz on the importance of plants to people – ask students to list 10 different ways in which we use plants or products from plants. Write the list on the board.
- Remind students that plants are multicellular organisms that can become very large indeed. Ask students to give levels of organisation within a multicellular organism and then read the sections headed 'Plant organ systems' and 'Plant organs'.
- Students complete **Worksheet 2.9.1**. They **EITHER** draw a plant, adding flowers, into their own notes **OR** draw a flower/flowers onto the plant and then label and annotate the leaves, stem, roots and flowers before sticking it into their own notes.
- Read through the final section on 'Plant tissues' with students, relating the tissues described in Fig 2.9 D to their function. Question students on how the position of each of the different tissues in the structure of the leaf is related to its function.
- Students draw and label Fig 2.9 D in their notes before completing questions 1–3 at the end of the spread.

Extension

Students complete **Worksheet 2.9.2** in which they describe the different levels of organisation in multicellular organisms and give both animal and plant examples.

Homework

Workbook page 24

Key words

shoot system, photosynthetic system, root system, xylem, phloem

2.10 Review answers

Student Book pages 60–61

Student Book answers

1. **a.** Organism; organ system; organ; tissue; cell; nucleus **[3]**
   b. plant organs: root and leaf; structure found in a plant cell: sap vacuole **[3]**
2. **a.** microscope **[1]**
   b. Looking at specimens too small to be seen with the eyes alone **[1]**
   c.

Part	Function
eyepiece	To look into the microscope and to magnify the specimen
objective lens	To magnify the specimen
stage	To put the specimen on so the light passes through it
coarse focus knob	To bring the specimen into focus
fine focus knob	To sharpen the image/improve the focus

[5]

3. **a.** *Nucleus* A, *cell wall* B, *cell membrane* C, *cytoplasm* D, *mitochondria* E, *sap vacuole* F, *chloroplast* G, *nucleus* H, *cytoplasm* J, *cell membrane* K and *mitochondria* L **[10]**
   b. Cell wall: supports the cell/protects the cell contents and gives the plant strength Sap vacuole: contains cell sap and presses against the cell wall to keep the cell firm and supported **[4]**
4. **a.** Specialised cell **[1]**
   b.

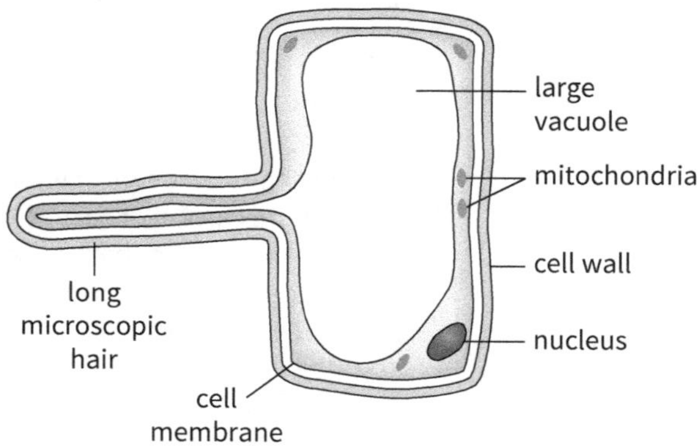

[5]

   c. They absorb water and mineral nutrients from the soil **[2]**
   d.

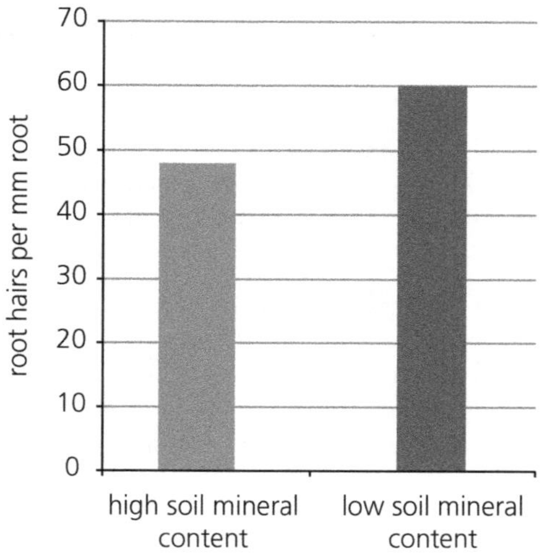

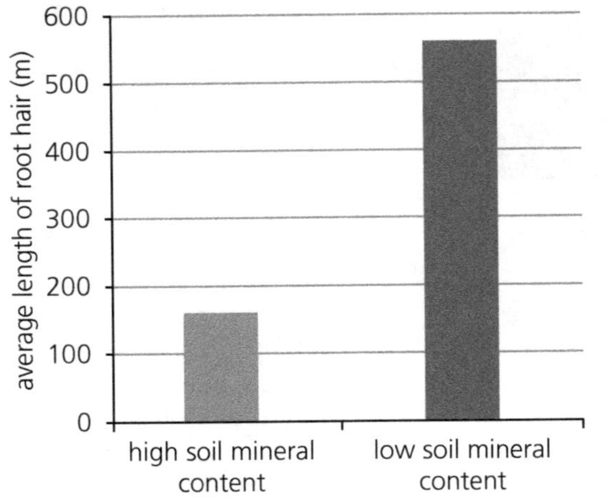

[5]

e. When there are not many minerals in the soil, plants grow more root hairs, and the root hairs grow a lot longer. They spread through the soil and absorb more minerals to supply the plant with what it needs. When there are plenty of minerals in the soil, the plant does not need as many root hairs and they are much shorter as they can absorb all the minerals they need from nearer the plant. Extra mark: the plant does not waste resources growing long root hairs when it does not need them. **[3]**

5. Neurone: carries electrical messages around the body

Root hair cell: absorbs water and mineral nutrients

Muscle cell: contracts to move parts of the body

Red blood cells: carries oxygen around the body

Fat cell: contains a fat store to provide fuel for respiration when it is needed

Palisade cell: carries out photosynthesis **[6]**

6. In the 17th century Hooke and Van Leeuwenhoek began to see cells using microscopes made from lenses they made and polished themselves. They both made careful drawings of the cells they saw. People did not believe them because it was very hard to make a good microscope at that time. Without a microscope you couldn't see cells. By the 19th century, there were many more microscopes around so many more scientists could observe cells for themselves. This meant they were happy to accept the idea that cells are the basic unit of all living organisms, because they had seen that every living organism they examined under the microscope was made up of cells. **[6]**

7. a. Total magnification = eyepiece lens magnification × objective lens magnification = 10 × 40 so *Euglena* is magnified ×400 **[3]**

b. Could be either as long as answer is supported by evidence, e.g., a plant – it has chloroplasts which trap light for photosynthesis, a process which happens in plants **[2]**

c. Similarities with a general animal cell – one from: *Euglena* has a nucleus, a cell membrane, no cell wall and a nucleus Differences – one from: chloroplasts, flagellum to move around, eye spot or contractile vacuole **[2]**

3.1 Microorganisms

Student Book pages 62–63

kerboodle

Online resources

- Worksheet 3.1.1 Growing microorganisms
- Worksheet 3.1.2 Modelling the growth of bacteria
- Interactive test

Extension

Students complete Q3 on page 63 calculating the magnification of bacterial cells.

Homework

Workbook page 25

Key words

microorganisms, single celled organisms, bacteria, fungi, viruses. culture, colonies, hyphae

Objectives

- Know that all organisms are made of cells
- Most microorganisms are single cells

Overview

This is the first of several lessons introducing students to microorganisms. It is very important both for CLSS success and for future studies that students recognise the positive importance of microorganisms in the natural world (e.g. in the cycling of nutrients in nature, in decomposition, on the skin and in the digestive system). The role of microorganisms in infectious disease is, of course, also important, but the way you introduce and work with students at this stage should give them a balanced overview of these vitally important organisms.

Activities

- Give each student three coloured pieces of card, e.g. green, red and yellow. Each colour represents an answer: green = true, red = false, yellow = I don't know. Read the following statements to students and ask them to raise the coloured card which best represents their own knowledge. It doesn't matter what they answer – the value of this exercise is for you to gauge how much students know about microorganisms and what misconceptions they already have. Note down the responses to each statement. Do NOT give students any feedback on their answers – simply tell them that they are now going to learn more about microorganisms. Repeat this exercise at the end of Unit 3.3 and see how the responses have changed. Suggested statements: *All microorganisms are so small you cannot see them. All diseases are caused by microorganisms. Microorganisms are bad for you. Microorganisms make food go bad. Some foods are made by microorganisms. Humans should aim to destroy all microorganisms.*
- Read through the content on bacteria on pages 62–63 with students. Discuss the features of bacteria and compare them with animal and plant cells that students are familiar with. Students can then answer Q2.
- Describe how quickly bacteria reproduce if they have everything they need. Ask students to suggest what bacteria might need to grow – food, water, right temperature. If you chose to culture bacteria on petri dishes with your students, to show how invisible microorganisms reproduce and form colonies, use **Worksheet 3.1.1**.
- Give students **Worksheet 3.1.2**. Students calculate the number of bacteria in a colony as it grows and then plot a graph of their data.
- Students read the rest of page 63, make notes on bacteria, fungi and viruses and answer questions 1, 4 and 5.

3.1 Student Book answers

1. Bacteria; fungi/yeasts; viruses.

2.

Feature	Animal cell	Plant cell	Bacterial cell
Size	(10–100) μm	(30–100) μm	(0.2–2.0) μm
Nucleus	Nucleus containing genetic material	Nucleus containing genetic material	No nucleus – single strand of genetic material
Cell wall	No cell wall	Cell wall made of cellulose	Cell wall but different structure to plants
Plasmids	No plasmids	No plasmids	Many bacteria have them

3. Actual length = measured length/magnification 7.4/74 000 = 0.0001 cm

4. Similarities: (one of) both are microorganisms; both are single cells; both have cell walls. Differences: (one of) yeast bigger than bacteria; different material in the cell walls; yeast have nuclei, bacteria do not.

5. plant cell: animal cell; bacterium; virus

Extension

3.2

Microorganisms are our friends

Student Book pages 64–65

kerboodle

Online resources

- Worksheet 3.2.1 Microorganisms at work: bread dough rising
- Worksheet 3.2.2 Poster planning
- Interactive test

Prior learning

- Know that all organisms are made of cells
- Most microorganisms are single cells

Extension

Complete your poster.

Homework

EITHER Complete your poster plan **OR** Workbook page 26.

Key words

mould, penicillin, antibiotic, fermenters

Objective

- Understand that microorganisms are typically single-celled organisms

Overview

In this extension lesson, students are encouraged to find out more about the ways in which we use microorganisms, extending their CLSS studies towards the understanding they will need to succeed in IGCSE Biology. Begin with familiar materials such as bread, yoghurt and cheese, and allow students to explore the role of microorganisms in their production, before considering less familiar uses of microorganisms such as in industrial processes to make medicines.

Activities

- Ask students to list ways in which people use microorganisms.
- Read through the first three paragraphs with students.
- If you are including practical work in this lesson, give students **Worksheet 3.2.1** and let them get their yeast mixtures started. Students can complete Q1 on page 65 while they are taking measurements.
- If you are not doing the practical worksheet, ask students to read through the section 'Microorganisms and food' and then answers questions 1 and 2.
- Read through the section on 'Microorganisms and medicine' with students, explaining difficult concepts. Explain how many people died of infectious diseases historically before you look at Figure 3.2 D with students and discuss the impact of antibiotics on human health. Make sure that students understand that antibiotics do not work against viruses.
- Students complete the page and answer questions 3 and 4.
- Give students **Worksheet 3.2.2** and ask them to complete the activity. Guide students to think about the age and understanding of the children they are making a poster for. If students complete their plans, they could make their posters which you could display or deliver to a local primary school for display.

3.2 Student Book answers

1. Yeast, sugar and water mixed → yeast feeds on sugar, respires and produces carbon dioxide → mixed with flour and water, the carbon dioxide makes dough rise → bread cooked – gas bubbles expand making bread lighter, yeast killed.
2. Yoghurt: bacteria mixed with warm milk; feed on sugar in milk; solidifies milk and produces lactic acid which gives sharp taste.
   Cheese: bacteria mixed with warm milk; feed on sugar in the milk; produce a lot of lactic acid, more than yoghurt bacteria; make milk separate into very solid curds and liquid whey; further bacteria may be added to change flavour over time.
3. Medicines that stop bacteria growing or kill them; used to cure bacterial diseases.
4. **a.** 1300
   b. accept 850 or 900
   c. 600
   d. 200

3.3 Microorganisms and disease

TWS

Student Book pages 66–67

kerboodle

Online resources

- Worksheet 3.3.1 Testing Pasteur's observations
- Interactive test

Prior learning

- Suggest and evaluate explanations for predictions using scientific knowledge and understanding and communicate these clearly to others.
- Say if and how evidence supports any prediction made

Extension

Students research and produce a timeline for the life and research of Louis Pasteur, with analysis of some of his major discoveries and some of the most challenging ethical aspects of his work.

Homework

Workbook page 27

Key words

germ theory of disease, vaccine

Objectives

- Describe how scientific hypotheses can be supported or contradicted by evidence from an enquiry
- Make a prediction on the likely outcome of a scientific enquiry based on your scientific knowledge and understanding

Overview

This lesson helps students to develop their skills in Thinking and working scientifically. Students are introduced to the world when we did not understand how infectious diseases are caused by microorganisms and had no effective treatments. They consider the work of Louis Pasteur, a global pioneer in germ theory and in vaccinations. Through the exciting story of Pasteur's anthrax trials, you allow students to make predictions of the outcome of the trials based on what they know, and help students understand how evidence is vital to support a scientific hypothesis, or to disprove it. This lesson gives you scope for setting project work on the life and work of Louis Pasteur.

Activities

- Find out from your students what they think causes infectious diseases and how they are spread. Then paint a picture of the 19th century CE, when no-one knew about the role of microorganisms in disease.
- Read the first paragraph with students and then give them **Worksheet 3.3.1**. Give your students a brief summary of the old idea of spontaneous generation of organisms from mud or dust, and then give them time to work through the questions, individually or as a class. If done individually, take time to go through the answers and explain the process and what was demonstrated by the swan-necked flasks.
- Read though the spread with students – perhaps ask different students to read sections. At the end of each section, pause and question students – for example, ask them to discuss Pasteur's ideas; would they have agreed to Rossignol's challenge? Do they agree with Pasteur's prediction? How did the evidence confirm Pasteur's hypothesis? Raise some of the issues of Pasteur's techniques with students. Think about the ethics of using an untried treatment in this way.
- Students complete questions 1 and 2.
- Q 3 is an opportunity for students to carry out independent research into Pasteur's work on rabies. Students can simply answer the question OR you could ask them to use their research into this gripping story to write a report on Pasteur and rabies for a TV news item, or for a series of posts on social media, or as a story for other students to read and understand both what happened, then and why Pasteur's work would not be possible today.

3.3 Student Book answers

1. Disease-causing germs are passed from one organism to another. In Pasteur's experiment, germs from infected animals were injected into his experimental animals. All of the unvaccinated animals developed the disease – confirming the idea of germs passed from one animal to another.
2. **a.** Using this account of Pasteur's famous experiment: e.g. he had a control group of animals which were not given the vaccine so he a comparison for his experimental group/he only changed one variable – whether the animals were vaccinated or not.

   b. E.g. he had not finished testing his vaccine before giving it to animals/he carried out an experiment that killed over 30 animals and in public with associated risk to public health/he lied about the vaccine he used.
3. Look for evidence of good research and a clear account of the story, including his motivation for the work, the good science in developing the vaccine and the bad science of using an untested vaccine on children.

3.4 Using science to prevent disease

SIC

Student Book pages 68–69

kerboodle

Online resources

- Worksheet 3.4.1 The impact of mask-wearing on deaths from COVID-19
- Worksheet 3.4.2 The impact of social distancing on COVID-19 infections
- Worksheet 3.4.3 Extension: Symptomless spreaders
- Interactive test

Prior learning

- Suggest and evaluate explanations for predictions using scientific knowledge and understanding and communicate these clearly to others.
- Say if and how evidence supports any prediction made

Objectives

- Describe how science is applied across societies and in research
- Describe how people develop and use scientific understanding

Overview

This lesson helps students to develop their understanding of Science in context. Any work relating to COVID-19 must be handled very sensitively as students may have been affected personally by the pandemic.

Pages 68–69 present students with information about the spread of infectious diseases caused by microorganisms. They consider a chart on the spread of infectious diseases produced by students BEFORE the COVID-19 pandemic of 2019 onwards, and look at how such scientific knowledge can be applied. You then present your students with evidence from two studies looking at the impact of both mask-wearing and social distancing on deaths globally and cases of COVID-19 in two Italian towns. Throughout this work, emphasise how scientific evidence changes understanding and so changes actions across societies.

Activities

- Any work relating to COVID-19 must be handled very sensitively as students may have been affected personally by the pandemic.
- Quiz students on how diseases are spread to see what they have retained from the previous lesson.
- Read through the opening paragraphs on page 68 and work through the display produced by the students. Make sure that students understand the different ways in which microorganisms can be passed from one person to another.
- Work through the section headed 'Using scientific knowledge' with students and encourage them to discuss the science behind the different instructions to prevent the spread of COVID-19. Students answer questions 1 and 2.
- The final section of page 69 challenges students to recognise that scientific theories must be backed by evidence. They are given two graphs to consider. **Worksheet 3.4.1** and **Worksheet 3.4.2** give students larger versions of the graphs on page 69, with extended questions. **EITHER** give students one or both of these worksheets to complete **OR** students work through the rest of the page, using the graphs to answer questions 3 and 4.

Extension

Give students **Worksheet 3.4.3**. This introduces the problem of people who have, or carry, a disease but have no symptoms themselves.

1. Students are asked to explain why asymptomatic disease is such a problem in controlling the COVID-19 pandemic.
2. If they have time, they are asked to investigate an historical case: Typhoid Mary, the story of Mary Mallon who is one of the first recorded cases of a symptomless person causing disease and even death in many other people.

Homework

Workbook page 28

Key words

droplet infection, direct contact

3.4 Student Book answers

1. COVID-19 had not appeared when the students did this work.
2. - *Wash your hands often and well and use antiseptic hand gels*: Your hands touch surfaces and can pick up viruses which you then get in your mouth/nose. Washing your hands/using antiseptic hand gel removes viruses/kills viruses, reducing your risk of disease.
   - *Wear a face mask to cover your mouth and nose when you go out*: If you cough or sneeze, droplets containing viruses are caught in your mask – reduces the risk of passing viruses to other people. Mask may also reduce the chance of you breathing in viruses.
   - *Meet as few people as possible – avoid crowded places*: People carry COVID-19. The more people you meet, and the more crowded the place, the more likely you are to be in contact with someone with the disease or – if you are infected – to pass it on.
   - *Keep a distance between yourself and other people and meet outdoors if possible*: The further away you are, the less likely you are to breathe in viruses from someone else. In the open air viruses and droplets are moved away more quickly [sunlight can kill viruses].
   - *Cough or sneeze into a tissue or into your elbow*: Any viruses you may have are not sprayed out into the air to infect other people.
   - *Stay at home away from other people if you feel ill:* You are most infectious when you are ill – keep away from other people so they don't come into contact with the virus.
3. In countries where people were made to wear masks outdoors and in shops quickly (pink line), the death rates from COVID-19 were very low. In countries (blue line) where there was more of a delay in making mask wearing important, there were more deaths. Countries where it took months to introduce mask wearing (green line) had a very high death rate.
4. Answer must be justified. Jada is correct – where businesses, etc., closed and people socially distanced/stayed at home, the number of people infected greatly reduced. That would in turn reduce the number of deaths. Where people carried on as normal [pink line] – the numbers infected just grew and grew.

3.5 The decomposers

Student Book pages 70–71

kerboodle

Online resources

- Worksheet 3.5.1 Classifying decomposers
- Worksheet 3.5.2 Observing decomposers
- Interactive test

Prior learning

- Sort living things into groups, using simple features

Objectives

- Describe the ecological role of some microorganisms as decomposers

Overview

In this lesson, you introduce students to the decomposers. They could set up their own specimens for mould to grow, or you can set some up in the days before the lesson (see **Worksheet 3.5.1**). This lesson prepares your students for future lessons on factors affecting the rate of decomposition and on the role of decomposers in food chains and webs.

Activities

- **EITHER** show students a piece of mouldy fruit **OR** ask them to imagine what would happen to food left in the fridge or outside for too long. Use a quick quiz/show of hands to find out which types of organisms students think act as decomposers.
- Read through the first three paragraphs on page 70 with your students and then do a thought experiment: ask students to estimate the mass of biological material produced locally in a year – the leaves, flowers and fruits of plants, grass that grows, droppings from animals, the dead bodies of animals and plants, human food waste, etc. Point out the value of the decomposers in breaking down this waste and restoring nutrients to the soil.

- Continue to read the next three paragraphs. Students answer questions 1 and 2.
- Students look at the Thinking and working scientifically box – which reminds them of the way in which scientists classify different organisms including decomposers. Students then use **Worksheet 3.5.1** to help them to complete Q3, making a simple key to identify the four different fungi shown in Fig 3.5 C, which are shown much larger on the worksheet.
- Give students **Worksheet 3.5.2**. There are two ways of using this worksheet. **EITHER** students set up the observation and do the first drawing or take a photograph on a smartphone, and then leave the specimen until the next lesson when they complete the practical **OR** provide students with specimens that are fresh and specimens that you have set up previously to go mouldy, so that they can make observations of them side by side.
- Students then read the rest of page 71 to help them to complete Q4 to reinforce their learning on the way humans use decomposers.

Extension

Students produce a flow chart to show the role of microorganisms in the production of compost. They then investigate the role of microorganisms in sewage treatment and also produce a flow chart for that process.

Homework

Workbook page 29

Key words

decomposers, compost

3.5 Student Book answers

1. A decomposer is an organism which breaks down the material in animal droppings, dead animals and plants, etc., and returns nutrients to the soil.
2. Prevents a build up of dead material and animal droppings; returns nutrients to the soil
3. Any useful key e.g.:

   KQ1 Is it growing on dead wood? Yes go to Q2 No go to Q3

   KQ2 Are the fruiting parts small and brown?

   Yes = beech bark canker

   No = jelly ear fungus

   KQ3 Are the fruiting parts purple?

   Yes = amethyst deceiver

   No = pin mould
4. Natural environment: fungi and bacteria grow on whatever is there – dead animal or plant, fruit, droppings, etc.; temperature varies with the weather, and by day and night; moisture levels not controlled; speed varies with where in the world it occurs/time of year; any other sensible points.

   Compost bin: only material from kitchen and garden put in; moisture could be controlled; decomposers confined with the waste; temperature can be partly controlled by colour of bin and using insulation material; any sensible points.

3.6 Investigating rotting rates

TWS

Student Book pages 72–73

kerboodle

Online resources

- Worksheet 3.6.1 Factors affecting decomposition rates
- Interactive test

Prior learning

- Use results to draw conclusions and to make further predictions

Extension

Write a sentence explaining what you would expect to see:

a. if Shakira kept a piece of potato at 40 °C

b. if Aardan kept a sample of milk at 50 °C.

Homework

Workbook page 30

Objectives

- Make conclusions by interpreting results, and explain the limits of the conclusions
- Evaluate investigations and suggest improvements, explaining your proposed changes

Overview

This lesson helps students to develop their skills in Thinking and working scientifically. It takes students through the process of developing a hypothesis, planning an investigation, collecting and analysing results and drawing conclusions. The questions encourage students to evaluate the results and suggest ways in which the investigation could be improved . They are provided with further data to practise their graphing skills.

Activities

- Look at Fig 3.6 A – ideally have the images on a big slide so everyone can see them without opening their Student Books. Talk about the information on decomposition rates in the first paragraph, relating it to the images.
- Students close their books – then ask them for ideas on how they might investigate the rate of decomposition. Questions might include:
  - o Different factors affect the rate of decomposition. Which would you choose to investigate?
  - o If you investigate the effect of temperature on decomposition, what is your hypothesis? Suggest how you might investigate this. What factors would you control? How would you measure decomposition? What do you predict would be the result of your investigation, based on your scientific knowledge?
- Students then work through pages 72–73 and answer questions 1, 2 and 3.
- When students have completed their answers, find out what improvements to the investigation they have suggested and use a show of hands to find out how many of the class have thought of the different ideas.

Ask students what happens to milk if it is left in a warm place. Most will have experience of milk going off. Give students **Worksheet 3.6.1** which looks at factors affecting the rate at which milk goes bad.

3.6 Student Book answers

1. They are infected by decomposers and digested away; they dry out.
2. She could have three pieces of potato at each temperature and take mean values – or similar.
3. She could observe what happens at lower temperatures, e.g. in a fridge/she could observe what happens at higher temperatures. Is there a maximum beyond which the decomposers do not work, for example?
4. a.

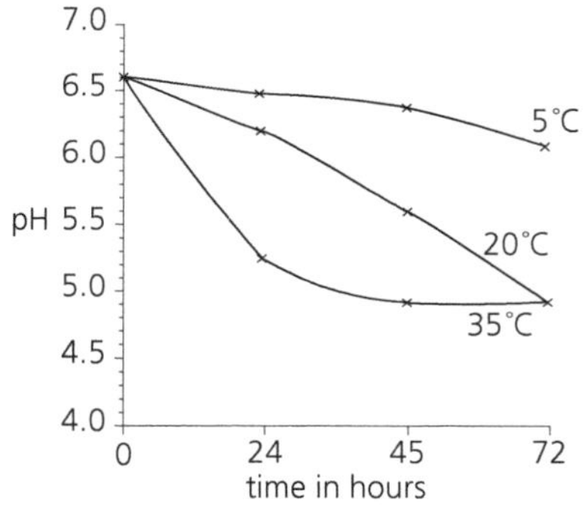

   b. Temperature increases the rate of milk decomposition. The rate at 5 °C is significantly slower than at 35 °C.

3.7 Food chains, food webs and decomposers

Student Book pages 74–75

kerboodle

Online resources

- Worksheet 3.7.1 Lion food web
- Worksheet 3.7.2 An Antarctic food web
- Worksheet 3.7.3 Extension: Fragile food webs in the Antarctic
- Worksheet 3.7.4 An African food chain
- Interactive test

Prior learning

- Know how food chains can be used to represent feeding relationships in a habitat and present these in text and diagrams
- Know that food chains begin with a plant (the producer) which uses energy from the Sun
- Understand the terms producer, consumer, predator and prey
- Explore and construct food chains in a particular habitat

Objectives

- Build and use food chains and webs including the importance of microorganisms as decomposers

Overview

In this lesson, you will revisit ideas of food chains that students studied in Stage 6, and build on those ideas to introduce the complexity of feeding relationships and the value of food webs as a model of these relationships. This gives the opportunity to reinforce the value of models to scientists and their strengths and limitations. Finally, you introduce the role of decomposers in food chains and webs.

Activities

- Before using the Student Book, ask your students what a food chain is. Make sure that they understand that food chains are simple models used by biologists to show the feeding relationships between organisms in a habitat. Students read through page 74 and answer Q1.
- Read through and discuss the Thinking and working scientifically box at the top of page 75 about food webs and go through Fig 3.7 C with students. The key concept here is that food chains are an over-simplification. Very few animals eat only one type of food. Food webs introduce a more realistic level of complexity – but students should understand that this is still simplified. Give students **Worksheet 3.7.1** to complete, which consolidates their understanding of both food chains and a food web.
- Once students have completed Worksheet 3.7.1, give them **Worksheet 3.7.2**. This introduces a very different model food web – the only plants are microscopic. The questions help students to recognise the interrelatedness of the organisms and how problems for one organism may cause problems for others. Ask students to consider the impact of the loss of other organisms within the web, e.g. the top predator (the leopard seals).
- Read through the final section on page 75 with students, drawing their work on decomposers into their understanding of food chains and webs. Give students **Worksheet 3.7.4**, which they complete by adding in the decomposers. Discuss the importance of decomposers in food webs and explain that they are rarely shown because it is difficult to draw them in.
- Summarise the key points of the lesson before students complete questions 3 and 4.

Extension

Students complete **Worksheet 3.7.3**, considering the interactions between organisms in a food web and the importance of keystone species.

Homework

Workbook page 31

Key words

producer, herbivore, carnivore, prey, predator, food web

3.7 Student Book answers

1. **a.** A simple model of the feeding relationships between organisms in a habitat.
   b. Any two valid food chains – only half marks if copied from the Student Book.
2. **a.** Food chains: advantages – one from: simple; easy to see how the energy flows.
   Disadvantage: over-simplifies as few organisms only have one food.
   b. Food webs: advantages – one from: give a better model of the real feeding relationships in a habitat; shows interdependencies between organisms.
   Disadvantages – one from: more complex and harder to work out; still an oversimplification.
   Any other sensible points.
3. Through respiration of the decomposers; respiration of plants which have absorbed the nutrients returned to the soil by the decomposers and used them in cells; respiration of animals which have eaten the plants which have absorbed the nutrients returned to the soil by the decomposers.
4. Look for awareness of inter-relatedness of organisms; importance of photosynthesis and producers in making food; flow of material and energy from plants to herbivores and carnivores; importance of decomposers in breaking down the material and returning nutrients to the soil so the chains and webs become cycles and the nutrients are restored for the next generation of plants, etc.
   Any other sensible points.

3.8 Review answers

Student Book pages 76–77

Student Book answers

1. **a.** A bacterium (allow bacteria); B virus/es; C fungus/fungi/yeast. **[3]**
   b. Any valid piece of information e.g. bacteria – no nucleus, may have plasmids; virus – nonliving, protein coat and genetic material; fungus – has a nucleus, cell wall different to plants. **[3]**
2. **a.** It has been attacked and digested by decomposers. They break down the apple and use it for respiration and growth, so the apple shrinks. **[3]**
   b. Any suitable food chain – must include at least 3 links and show the role of the decomposers in the environment. **[5]**
3. **a.** Bacteria feed on the sugar in milk and break it down into lactic acid. This solidifies the milk and gives it a sharp taste. **[2]**
   b. She should control the quantity of milk/ and bacteria/ and the containers they are in. She should only change the temperature. Any other sensible point. **[2]**
   c. When the pH is 4.5 (pH of yoghurt). **[1]**
   d. **[4]**

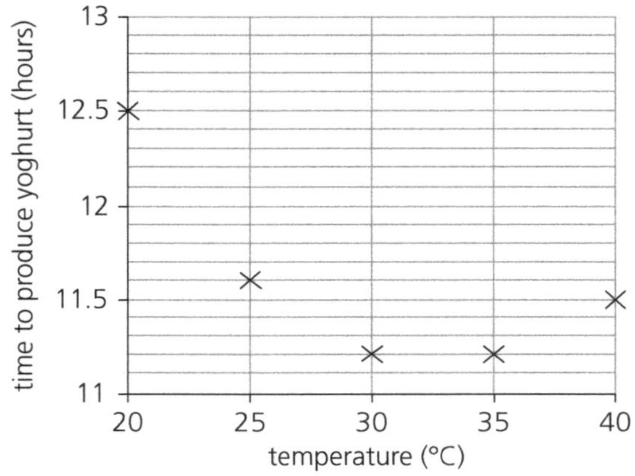

   e. 32.5 °C **[1]**

4. a. When the dough with yeast and sugar is kept in a warm place, it gets bigger. Dough without yeast does not change size. **[2]**

b. Yeast uses the sugar to respire. It produces carbon dioxide gas as a waste product. The bubbles of gas make the dough rise and expand. **[2]**

c. Cylinder with yeast: increase the temperature it is kept at. Cylinder without yeast – add yeast. **[2]**

5. a. Decomposition **[1]**

b. Microorganisms/decomposers **[1]**

c. Fungi, bacteria **[2]**

d. The actions of the decomposers break down the dead bodies of animals and plants, and animal droppings, etc. They return the nutrients to the soil to be used again by plants, and then passed on to animals. Also prevents build up of dead material, etc. Any other sensible point. **[3]**

6. a. Any three from droplet infection, contact, contaminated food and drink and breaks in the skin. **[3]**

b. Any three (with explanation) from:

- Wash your hands often and well and use antiseptic hand gels: To kill microorganisms on your hands and prevent you transferring them to your mouth or other people.
- Wear a face mask to cover your mouth and nose when you go out: to prevent droplets containing microorganisms going from your mouth into the air for others to breathe in/ also prevent you breathing in large droplets from others.
- Meet as few people as possible – avoid crowded places: The fewer people you meet, the lower the risk of meeting someone who is carrying and spreading disease-causing microorganisms.
- Keep at a distance from people and meet outdoors if possible: the further you are from people, the lower the risk of spreading microorganisms between you. Outside conditions spread microorganisms and may kill them.
- Cough or sneeze into a tissue or into your elbow: minimises the spread of droplets carrying microorganisms.
- Stay at home away from other people if you feel ill: reduces the risk of spreading the microorganisms that are making you ill **[6]**

7. a. The effect of temperature on decomposition – they use two different temperatures. The effect of microorganisms – they sterilised some of the leaves so they had no microorganisms on them. **[4]**

b. To make it a fair test – the leaves of different types of trees could decompose at different rates. **[1]**

c. To make it a fair test – so they could compare the results from the different groups. **[1]**

d. Any answer which is fully justified. Best answer would be: Group F. These leaves are not washed in sterilising liquid so they will have lots of decomposers on their surface, and they are kept at a warm temperature which allows the decomposers to digest the leaf material quickly. They will lose more mass through evaporation of water than the leaves in the sealed container. **[5]**

4.1 The physical environment

Student Book pages 78–79

kerboodle

Online resources

- Worksheet 4.1.1 The physical environment crossword 1
- Worksheet 4.1.2 The physical environment crossword 2
- Interactive test

Prior learning

- Plant growth is affected by temperature and water

Extension

Students investigate the advantages and disadvantages for organisms of living and breathing in the water rather than on land.

Homework

Workbook page 32

Key words

solvent, insulating

Objective

- Describe the importance of water and the water cycle

Overview

This short chapter covers water in the environment and the water cycle. This is students' first introduction to the interactions between the environment and living organisms. If students develop a sound understanding of the interactions of abiotic and biotic factors in the environment, this will support their success in future at IGCSE. In this lesson, you help your students to understand why water is so important for the success of life on Earth.

Activities

- Introduce the topic of water and life – then ask students to estimate how much water is in the human body and note it down.
- Question students on the non-living factors that affect living organisms in the natural world. Begin with things that are familiar to them such as temperature, rainfall, etc. Then read through the first two sections on page 78. Students answer questions 1 and 2 on page 79.
- Now get students to think about the importance of water. Find out how much water they thought was in the human body – ask for ideas or start with everyone standing up and ask anyone who didn't put 70–72% to sit down. See how many students got it right!
- Read through the section on the importance of water and look at the water content of different human organs (which also revises knowledge of organs in the body). Students answer Q3 which asks them to draw a bar chart.
- Talk about the properties of water and demonstrate as many of them as possible. For example:
  1. Have a number of different chemicals that will dissolve in water – e.g. salt, sugar, copper sulphate, potassium manganate (VII) – and dissolve them to demonstrate that water is a good solvent.
  2. Put ice cubes in water so students can see that ice floats.
- Students answer Q4 on page 79.
- Finally, students complete a crossword puzzle to test their knowledge and understanding of this lesson. **Worksheet 4.1.1** has the answer choices available; **Worksheet 4.1.2** does not. The questions are the same on both.

4.1 Student Book answers

1. Any two from rocks/soil/sand/temperature/availability of water/any other sensible point.
2. Rocky desert: little fertile soil; little water available; a lot of light. Tropical rainforest: lots of fertile soil; lots of water available; light depends on where in the forest the plant is growing.
3. a.

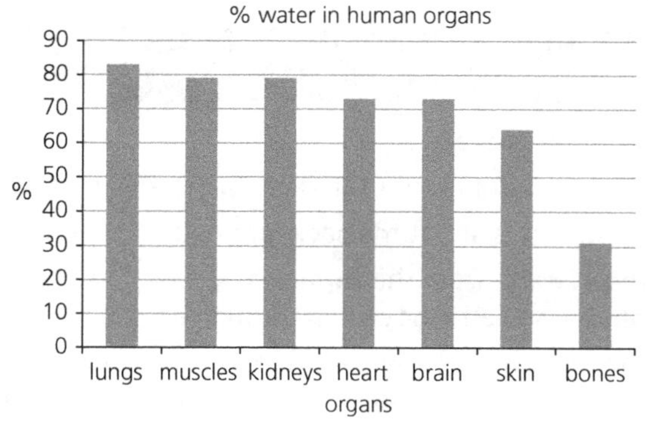

   b. The lungs
   c. The bones
4. Good solvent: many things dissolve in it so it is useful for transport/chemical reactions take place in it; any other sensible point.

   Temperature remains relatively stable/takes a lot of energy to heat up and cools slowly: gives a very stable environment for organisms – they don't have to cope with big changes in temperature.

   Ice floats on water and insulates it: prevents water underneath freezing and killing the organisms living in it.

4.2 The water cycle

Student Book pages 80–81

kerboodle

Online resources

- Worksheet 4.2.1 Ice to water vapour: from solid to liquid to gas
- Worksheet 4.2.2 Water cycle poster
- Interactive test

Extension

Complete your poster.

Homework

Workbook page 33

Key words

water cycle, evaporation, condensation, precipitation, collection, open water, groundwater, runoff, transpiration

Objective

- Describe the water cycle

Overview

In this lesson, your students will learn about the water cycle. You can provide them with the opportunity to see the events of the water cycle in action in a simple practical that extends over at least one lesson. Students should make posters or mobiles to illustrate the events of the water cycle to reinforce knowledge and understanding of this fundamental process. Not only is it important biologically, but it is also being affected by climate change – something students will consider later in this course and in IGCSE Biology.

Activities

- Read the first paragraph on page 80 to students. The idea that the water we drink has passed through living organisms since the earliest times – dinosaurs are always a great example – and that the great figures in our history used the same water that we now drink is one that students usually find amazing!
- Ask students where we find water. Build up a list on the board/screen of as many ideas as possible – you may find your students have lots of ideas or none! Try to end up with a list that stretches from clouds/atmosphere through glaciers and rivers and oceans to puddles, oases, reservoirs, animals and plants. The more ideas students have, the better.
- Remind students of the three states of water – coordinate with their chemistry teacher to make sure this has been covered in their chemistry lessons.
- Read through the text on the water cycle with students – stop to discuss each bullet point and question students to make sure they understand. For example, the concept of evaporation when the water is NOT boiling can be difficult for students to grasp.
- Give students **Worksheet 4.2.1**. This is a relatively simple practical that needs little equipment – you can adapt it for use without a thermometer or balance, if necessary, although the combination given makes the best experience for students. This shows them the stages of the water cycle from ice to liquid water and then the loss of water by evaporation over time. If you choose, you can leave the beakers of water out in the laboratory/classroom until all of the water has evaporated away!
- Students write up their investigation, leaving space to complete their results and discussion after the next lesson. Then students answer questions 1 and 2.
- Students should **EITHER** draw a water cycle in their books **OR** complete **Worksheet 4.2.2**, which gives them instructions for making a poster of the water cycle. You need to provide resources for these activities. Posters can be displayed in your classroom.

4.2 Student Book answers

1. a. Water as a gas/in a gas state.
   b. The process by which liquid water turns into a gas as it gets warmer.
   c. When water vapour cools and turns back into liquid water.
   d. The evaporation of water from the surface of leaves. The water has been taken from the soil and moved up through a plant.
2. a. Open water → heating by sun → evaporation → water vapour in atmosphere → cooling → condensation of water vapour to form water droplets (clouds) → precipitation – river/open water/groundwater → water purification → taken to homes → Rasheed's cup.
   b. Thousands of years.

4.3 Global warming and the water cycle

SIC

Student Book pages 82–83

kerboodle

Online resources

- Worksheet 4.3.1 The effect of temperature on the rate of ice melting
- Worksheet 4.3.2 The effect of temperature on the evaporation rate of water
- Worksheet 4.3.3 Raja's evidence
- Worksheet 4.3.4 Mariam's evidence
- Interactive test

Prior learning

- Suggest and evaluate explanations for predictions using scientific knowledge and understanding and communicate these clearly to others.

Extension

Students answer Q5.

Homework

Workbook page 34

Objectives

- Describe the water cycle
- Discuss how the uses of science can have a global environmental impact

Overview

This lesson helps students to develop their understanding of Science in context. Students are given a brief overview of global warming and its impact on the water cycle and they will evaluate evidence for this. They will learn much more about global warming and climate change in later spreads – here they simply need to be aware that it is happening. Worksheets provide two practical activities in which students can see and measure the impact of temperature increases on features of the water cycle – ice melting and water evaporating – and handle data extracted from the main content of the spread.

Activities

- Introduce students to the main concepts of global warming as given in the first paragraph of page 82.
- Ask students to summarise the water cycle and then suggest ways in which a rise in global temperatures might affect the events of the water cycle.
- Give students **Worksheet 4.3.1**. In this activity, students set up beakers of ice at different temperatures and observe and record the speed at which the ice melts. They can plot a simple graph of their results.
- Once students have set up their ice beakers, give them **Worksheet 4.3.2**. Students set up beakers of water at different temperatures and leave them for a week to observe the different amounts of water lost, before showing their results as a bar graph.
- In both experiments, encourage students to think about what they are observing in the context of global warming and the effect of temperature increases on the water cycle. Students complete questions 1 and 2 on page 83.
- Ask half of your students to look at Raja's research and answer Q3. A larger version of the graph is available on **Worksheet 4.3.3**.
- The other half of the class look at Mariam's research and answer Q4. A larger version of the graph is available on **Worksheet 4.3.4**. Depending on the time available, **EITHER** different students report back to the class on their findings **OR** the groups swap over and look at the other piece of evidence.

4.3 Student Book answers

1. Scientific knowledge used to develop petrol engines and ways of using fossil fuels to generate electricity. Burning fossil fuels produces waste carbon dioxide gas into the atmosphere. Carbon dioxide is a greenhouse gas. Increased levels in the atmosphere are warming the whole planet.
2. 2.5% of the water on Earth is fresh water.

   68.7% of the fresh water on Earth is stored as ice.
3. 17 cm
4. e.g. Australia and parts of the US are getting 50% less than normal; Middle East/north Africa/Greenland are getting 50% more rain than normal.
5. **a.** Temperature of the Earth increasing/ice/glaciers melting causing rise in sea level/sea temperature increasing so water expanding causing rise in sea levels. Lack of rainfall in some areas resulting in droughts. Extreme rainfall in some areas leading to flooding. Any other sensible points.

   b. Any of these problems may be chosen as long as there is clear justification for the choice.

4.4 Review answers

Student Book pages 84–85

Student Book answers

1. a. A non-living factor in the environment. [1]
   b. Any two from: rocks/soil/sand; temperature; water availability; any other sensible point. [2]
2. a. A good solvent means many things dissolve in it. Important to living things because it carries substances around the body, reactions of life take place in it. [3]
   b. The ice forms an insulating layer on the top of the water, so the water underneath it does not freeze. The animals and plants living in this water do not freeze and die. [2]
3. a. [8]

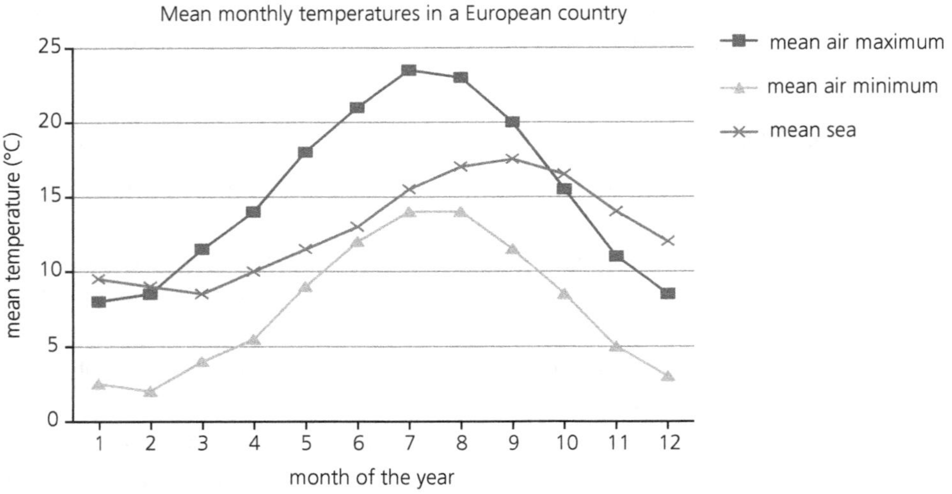

   b. Mean sea temperatures are lower than air temperatures all year.
   Range of mean sea temperatures is smaller than range of air temperatures, e.g. min 8.5 °C, max 17.5 °C, whereas minimum air temp is 2.0 °C and maximum air temperature is 23.5 °C. Any other sensible points. [3]
   c. Sea organisms do not have extremes of temperature to deal with – 9 °C difference all year – compared with land organisms, so they have less temperature stress, which makes life easier.
4. a. Data shows a trend of rising sea levels over time; by 17 cm over 100 years. [2]
   b. Melting ice from glaciers, etc.; sea water temperatures are increasing so water is expanding. [5]
5. a. A = evaporation, B = transpiration, C = condensation, D = precipitation, E = collection. [5]
   b. A Evaporation: Energy from the sun heats up the water in puddles, streams, lakes, rivers and oceans. Some of the water evaporates. It turns from a liquid to a gas called water vapour. [3]
   B Transpiration: The evaporation of water from the leaves of plants. [1]
   C Condensation: Water vapour rises up into the sky. The higher it gets, the cooler it is. When the water vapour cools down, it turns back into liquid water. This is condensation and it forms clouds in the sky. [3]
   D Precipitation: The water droplets in the clouds get bigger and heavier as more water condenses. When they get big enough, they fall back down to Earth as precipitation. Precipitation may be rain, snow, hail or sleet, depending on the temperature of the air. [3]
   E Collection: Precipitation collects in bodies of water, and the whole cycle starts again. It may fall directly into open water such as oceans, lakes and rivers. The precipitation may soak through the ground until it reaches a layer it cannot travel through, then be stored there. Water held in the soil and stored underground is called groundwater. Water that cannot be absorbed by the ground, but runs off the surface into streams and rivers, is called run-off. [4]
   c. X = run-off; Y = open water [2]
   d. Ice/glaciers; ground water in the soil and underwater stores. [2]

Stage 7

Review answers

Student Book answers

1. **a.** Sensitivity, movement, nutrition/feeding **[3]**
   b. respiration, reproduction, excretion, growth **[4]**
2. **a.** **i.** E **ii.** B **iii.** G **iv.** F **[1, 1, 1, 1]**
   b. Chloroplast; it traps light to make food for the plant. **[2]**
   c. Nucleus, cell membrane, cytoplasm. **[3]**
3. Reproductive system = produces new organisms
   Digestive system = breaks down food so it can be absorbed into the body
   Respiratory system = takes in oxygen and removes carbon dioxide
   Circulatory system = transports materials around the body **[4]**
4. **a.** B – has a root hair which gives a big surface area for the uptake of water and mineral salts from the soil. **[2]**
   b. C – cells have many chloroplasts for capturing sunlight. **[2]**
   c. Red blood cell – three ways from: has no nucleus to make more space to carry oxygen; has haemoglobin to carry oxygen; is very small and flexible to squeeze through tiny blood vessels; biconcave to give a bigger surface area to take in oxygen. **[4]**
5. **a.** Clear diagram with labels showing cell wall, genetic material, plasmids, cytoplasm as in Fig 3.1 B on page 62 of the Student Book. **[5]**
   b. 3 from: droplet infection; direct contact; contaminated food and drink; breaks in the skin. **[3]**
   c. Hand washing removes microorganisms from the skin – washing using water alone reduces the percentage of microorganisms on the hands by 21% (from 44% to 23%); using soap and water is better, it reduces organisms on skin by 36% down to 8%. **[3]**
   d. Any 2 from: wearing a face mask; avoiding crowded places; coughing into tissue or elbow; stay at home if you feel ill; avoid crowds. **[2]**
6. **a.** Decomposition **[1]**
   b. Bacteria, fungi **[2]**
   c. **i.** Brown substance full of nutrients produced when decomposers break down garden and kitchen waste. **[1]**
   ii. Make it warmer e.g. dark colour bin, position in the sun or insulate. **[1]**
7. Clear diagram with labels showing evaporation, condensation, precipitation and collection as in Fig 4.2.B. **[5]**
   Extra 2 marks for labelling open and ground water and run-off. **[2]**

8. a. Exemplar – there are other ways of doing a key which would also work.

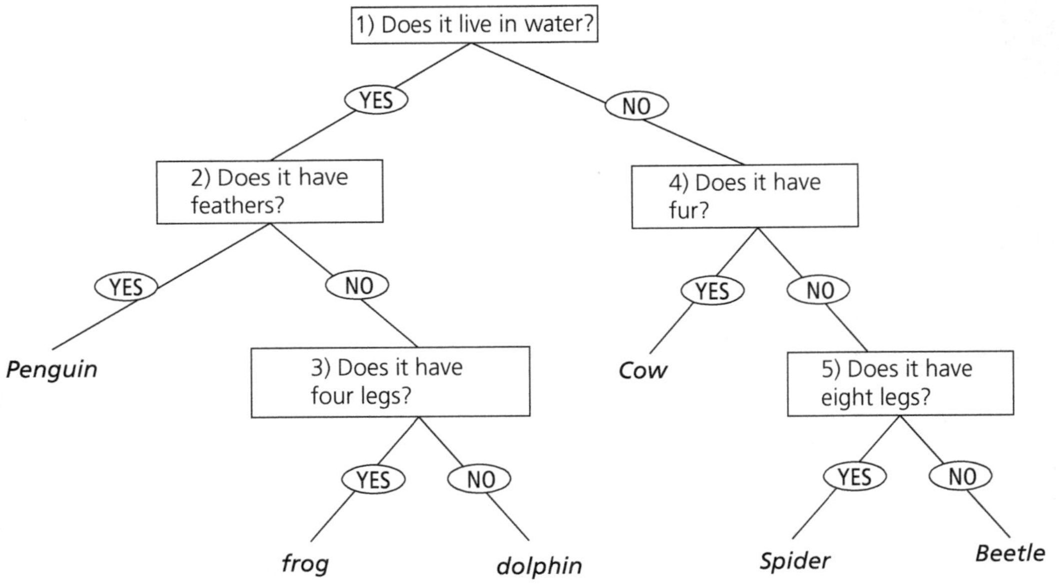

[6]

b. i. A, B. C and D or cow, dolphin, penguin, frog **[4]**

ii. Animals with backbones; plus one of: often relatively large or most live on land **[2]**

iii. Cow and dolphin are mammals **[1]**

3 from: **[3]**

- They all have fur
- They give birth to live young and make milk to feed them.
- They have four limbs.
- They breathe air using lungs.
- Control their own body temperature.

c. So that all scientists, wherever they are, know what organism is being identified. **[1]**

5.1 Diffusion in biology

Student Book pages 88–89

kerboodle

Online resource

- Worksheet 5.1.1 Modelling diffusion
- Interactive test

Extension

Students chose one example of diffusion in the natural world and explain how it works through a series of annotated diagrams.

Homework

Workbook page 35

Key words

diffusion, net, concentration gradient

Objectives

- Describe how the structure of the human respiratory system is related to its function of gas exchange
- Describe the diffusion of oxygen and carbon dioxide between the blood and the lungs

Overview

This spread prepares the ground for work on the respiratory system and gas exchange by introducing students to the concept of diffusion in the context of biology (they may have met it in chemistry so it is helpful to discuss this with their chemistry teacher). Diffusion is a key concept in many biological systems both for CLSS and at IGCSE, so making sure that students have a basic understanding of the principles now is time well spent. It is important to avoid misconceptions about diffusion – for example, the movement of the particles is RANDOM – they do not deliberately move to areas of high concentration.

Activities

- Begin the lesson with an experiment. Stand in front of your class and tell students that you are going to squirt perfume/aftershave/air freshener somewhere in the class. Students should all close their eyes and raise their hand when they can smell the perfume. Once they have raised their hands, they can open their eyes and watch what happens with the rest of the class. This is a practical demonstration of diffusion.
- Question students to get an understanding of how well they understand diffusion from their chemistry lessons if appropriate.
- Set up an experiment on your desk – take a beaker of water and drop a crystal of potassium manganate (VII) into it. Leave it for diffusion to take place throughout the lesson. Look at it with students and discuss what it happening several times throughout the lesson.
- Read through page 88 with students and discuss the process. Emphasise that the movement of the particles is random but that the particles will always end up spreading out from where there are a lot of them to where there are fewer. Make sure that students understand that diffusion only takes place in fluids – liquids and gases.
- Read page 89 with students who then answer questions 1–3.
- Students complete **Worksheet 5.1.1**. This is a Thinking and working scientifically activity – students draw diagrams to show the particles spreading by diffusion in the beaker of water and potassium manganate (VII). They then evaluate the model, discussing its strengths (e.g. it helps you to visualise what is happening when the particles are too small to see) and weaknesses (e.g. what fills the space between the particles/the particles are too big and there aren't enough of them, etc.).
- Students attempt Q4 and keep their answers for later in this series of lessons.

5.1 Student Book answers

1. Matter is made up of moving particles which are too small for us to see.
2. The net movement of particles down a concentration gradient, from a high concentration to a low concentration.
3. There is a high concentration of scent in the flowers. The scent particles spread through the air by diffusion. Eventually, some of the scent particles will reach your nose and you will smell them.
4. High concentration of oxygen in the air, low concentration of oxygen in your blood; oxygen moves by diffusion down the concentration gradient from the air to the blood.

5.2 Aerobic respiration in animals and plants

Student Book pages 90–91

kerboodle

Online resources

- Worksheet 5.2.1 The word summary for aerobic respiration
- Worksheet 5.2.2 [EXT] Mitochondria and energy needs
- Interactive test

Prior learning

- Identify and describe the functions of mitochondria
- Describe the seven characteristics of living organisms

Extension

Students answer **Worksheet 5.2.2.** Students should show awareness of cells that use a lot of energy, e.g. muscle cells, heart muscle cells, cells in growing regions of bones or skin of young children, growing shoots of plants, forming fruits and tissues/cells with low energy demands, e.g. fat cells, starch storage cells in plant roots or stems, areas where little growth takes place, ageing tissue. Give extra credit where students include plant examples.

Homework

Workbook page 36

Key words

oxygen, glucose, aerobic respiration, carbon dioxide, water

Objectives

- Know that aerobic respiration occurs in the mitochondria of plant and animal cells, and gives a controlled release of energy
- Know and use the summary word equation for aerobic respiration

Overview

In this lesson, you explore the process of aerobic respiration with students. This is an excellent opportunity to make sure that your students have remembered and understood work on the characteristics of living organisms and the functions of cell structures from Stage 7 Biology. They apply and extend this knowledge and understanding as you explain the importance of the process of aerobic respiration in cells. Understanding cellular respiration is key both for understanding the need for gas exchange systems and for later study in IGCSE Biology.

Activities

- Look at the images in Fig 5.2 A on page 90 of the Student Book. Ask students to name other examples of living organisms and ask for the seven characteristics of life.
- Explain to students that cells need energy to do work – from muscle cells contracting to building the chemicals needed for the body to function. Demonstrate burning sugar in air, using appropriate safety precautions, or show a video clip. This is uncontrolled energy release. Discuss with students the need for cells to release energy in a controlled way.
- Read through the section on aerobic respiration. Write the word summary on the board and talk about all of the substances with your students: glucose, oxygen, carbon dioxide and water. Make sure that students know what each of them is – a sugar, a gas from the air, etc., and where it comes from. They should also learn that carbon dioxide is toxic and must be removed from the cells.
- Explain to students that the substances involved in aerobic respiration move into and out of the cells by diffusion from the blood.
- Give students **Worksheet 5.2.1**. Students produce a word summary of respiration. They should make this large and clear and decorate it in some way to make it easier to remember. They can stick this in their notebooks or you could make a wall display for the classroom.
- Ask students to name the part of a cell where aerobic respiration takes place – some will remember the mitochondria. Remind students that both animal and plant cells have mitochondria. Students read through the text on page 91 and answer questions 1–4.

5.2 Student Book answers

1. Aerobic respiration
2. Mitochondria
3. glucose + oxygen → carbon dioxide + water + energy
4. Your heart is always beating so heart muscle works hard all the time. It needs a lot of energy. Mitochondria supply energy from aerobic respiration, so heart muscle has many mitochondria.
5. Aerobic respiration uses oxygen to release energy from food in a controlled way. All cells, both in animals and plants, need this energy to carry out the processes of life such as growth, movement, reproduction, excretion, repair, etc. so the same process takes place in both types of cells.

Extension 5.3 Anaerobic respiration

Student Book pages 92–93

kerboodle

Online resources

- Worksheet 5.3.1 How easily do finger muscles tire?
- Worksheet 5.3.2 Does training change your performance?
- Worksheet 5.3.3 Specialised muscle cells
- Interactive test

Objectives

- Know that aerobic respiration occurs in the mitochondria of plant and animal cells, and gives a controlled release of energy
- Know and use the summary word equation for aerobic respiration

Overview

In this Extension lesson, you will be reinforcing what students have learned about aerobic respiration in Unit 5.2 and introducing them to the key principles about anaerobic respiration that they will need to understand for IGCSE Biology. These include:

Defining *anaerobic respiration* as the chemical reactions in cells that break down nutrient molecules and release energy for metabolism without using oxygen.

Stating the word equation for anaerobic respiration in muscles during vigorous exercise (glucose → lactic acid).

Knowing that anaerobic respiration releases much less energy per glucose molecule than aerobic respiration.

Activities

- Begin by asking students to tell you the word summary equation for aerobic respiration. Remind them that cells need plenty of oxygen to carry out aerobic respiration and ask students how the oxygen gets to the cells (in the blood).
- Students read through the section titled 'Respiration without oxygen'. Discuss the problems of respiring without oxygen and the production of lactic acid. Help students to understand that breaking down many glucose molecules provides the body with the energy it needs when there is not enough oxygen for aerobic respiration, but this is inefficient.
- Students work through **Worksheet 5.3.1**. In this simple exercise, students feel their muscles tire and ache a little as they are forced to respire anaerobically. Students use their results to draw a graph which demonstrates that the muscles continue to work for some time without oxygen, by respiring anaerobically, but eventually they will not contract any more.
- Students read page 93 on lactic acid. Explain how lactic acid builds up in the muscles during anaerobic respiration and how the body needs more oxygen once exercise stops to break down the lactic acid. Discuss the benefits of training in developing a better blood supply to the muscles – this delivers more oxygen so it takes longer for anaerobic respiration to be required, and it gives a faster recovery as more oxygen reaches muscles to break down lactic acid once exercise stops.
- If students are interested, give them **Worksheet 5.3.2** to do at home to see if they can improve their performance in the peg test over time.
- Students answer questions 1–5 on page 93.

Extension

Students work through **Worksheet 5.3.3** looking at different types of muscle cells, the respiration they carry out and their roles in the body.

Homework

Workbook page 37

Key words

anaerobic respiration, lactic acid

5.3 Student Book answers

1. Anaerobic respiration
2. e.g.

Aerobic respiration	Anaerobic respiration
needs oxygen	does not need oxygen
waste product: carbon dioxide	waste product: lactic acid
a lot of energy per molecule of glucose	less energy per molecule of glucose
happens all the time	only happens for short bursts, e.g. when exercising hard

3. Sometimes cells do not have enough oxygen to carry out aerobic respiration but they still need energy. Anaerobic respiration makes it possible to release some energy from food even with very little or no oxygen.
4. When Amir is running, his muscles do not get enough oxygen and use anaerobic respiration. This causes a build-up of lactic acid. Lactic acid is the waste product of anaerobic respiration. When he stops running, his body needs extra oxygen to break down the lactic acid in his muscles. He keeps breathing faster for several minutes to supply the extra oxygen needed to break down the lactic acid which built up when he was running OR Once the lactic acid has been broken down, Amir's breathing returns to normal.
5. Red part: anaerobic respiration makes lactic acid; the longer the exercise continues, the more lactic acid builds up.
   Blue part: during recovery, extra oxygen is used to break down and remove the lactic acid.

5.4 Investigating respiration

TWS

Student Book pages 94–95

kerboodle

Online resource

- Worksheet 5.4.1 Is there more carbon dioxide in the air I breathe out than in the air I breathe in?
- Interactive test

Objectives

- Make predictions of the likely outcome of a scientific enquiry based on your scientific knowledge and understanding
- Plan an investigation, considering the variables involved
- Know the meaning of hazard symbols

Overview

This lesson helps students to develop their skills in Thinking and working scientifically.

It provides examples of three different investigations using the presence of carbon dioxide to demonstrate that aerobic respiration has taken place. Students are asked to evaluate the different methods, looking for strengths and weaknesses in the experimental designs.

Activities

- Begin by asking students to remember the equation for aerobic respiration. Talk to them about the use of indicators in science and explain that the change in colour of limewater indicates the presence of carbon dioxide. Discuss the important features to look for in investigations.
- Students read through to the end of investigation 1. Demonstrate this – breathe out gently through a straw to show students how limewater turns from clear to cloudy. Comment on the hazard symbol on the bottle of limewater and remind students of the importance of safety in practical work. Alternatively, give students the opportunity to try this for themselves.

- You could also demonstrate that if you keep breathing through limewater it eventually turns clear again – one weakness in the investigation.
- Students read through investigations 2 and 3 before answering questions 1–3.
- Have a plenary session asking students for strengths and weaknesses of the different investigations that they have identified.
- Do a tally on which investigation students think is the most useful to a biology teacher, asking them for reasons. If more than one investigation gets votes, get students to draw a pie chart to show their conclusions with reasons given for each choice.
- If there is time, give students **Worksheet 5.4.1** which challenges them to either carry out an investigation into their own carbon dioxide production or asks them to explain how a piece of apparatus might be used to do that.

Extension

Students plan an investigation to show that people breathe out more carbon dioxide after exercise that they do at rest.

Homework

Workbook page 38

5.4 Student Book answers

1. Investigation 1: Whether people have carbon dioxide in the air they breathe out.

   Investigation 2: Whether the air people breathe out contains more carbon dioxide than the air they breathe in.

   Investigation 3: Whether germinating seeds respire and produce more carbon dioxide than seeds that are dormant.
2. Investigation 1: Good points: simple; easy to carry out; easy for students to see what happens.

   Problems: no control; no clear comparison; different amount of air from each person's lungs – limited in value; any other sensible points.

   Investigation 2: Good points: relatively simple and clear to carry out; good control; makes students think about the process which causes the difference between inhaled and exhaled air.

   Problems: the volume of air bubbled through the limewater is different both for each person and between the lungs and the air in a syringe or bottle; any other sensible point.

   Investigation 3: Good points: well thought out; valuable to reinforce the concept of respiration occurring in all living organisms; two controls – one with no seeds and one with seeds that are not growing – so students can see that it is not the presence of seeds that is affecting the limewater.

   Problems: It takes time to complete – at least a week. It does not show the students anything about respiration in humans; any other sensible point.
3. **EITHER** Investigation 2: Shows students the difference in the carbon dioxide concentration in inhaled and exhaled air; leading to discussion/understanding of the processes of respiration and gas exchange. Can be used for further investigations into the effect of exercise on the amount of carbon dioxide produced, etc. Any other sensible justification.

   OR Investigation 3: Shows students respiration in plants with clear demonstration that actively growing plants respire and produce more carbon dioxide than seeds which are not growing. Valuable for emphasising the importance of respiration in different types of organisms.

   OR both of these investigations for the reasons given above.

5.5 The lungs and gas exchange

Student Book pages 96–97

kerboodle

Online resources

- Worksheet 5.5.1 The human respiratory system
- Worksheet 5.5.2 The parts and functions of the human respiratory system
- Worksheet 5.5.3 The human respiratory system – what do you know?
- Interactive test

Prior learning

- Use scientific names and position of some of the major organs in the body
- Describe the main functions of some of the major organs and explain how they are essential

Extension

Complete **Worksheet 5.5.3**.

Homework

Complete **Worksheet 5.5.3** or Workbook page 39.

Key words

respiratory system, gas exchange, gas exchange system, mouth, nose, trachea, bronchus (bronchi), alveolus (alveoli), pleura, diaphragm

Objectives

- Describe the structure of the human respiratory system
- Relate the structure of the human respiratory system to its functions

Overview

In this lesson, you move your students away from aerobic and anaerobic respiration taking place in the cells and introduce the major organ system of the lungs. Students commonly get very confused by respiration and breathing. This isn't helped by the term 'human respiratory system'! It is helpful to talk about the **gas exchange system** instead. This lesson is a great opportunity to make sure that your students are clear about the difference. If you can avoid misconceptions and confusion during their CLSS course, it will help students do well later in their IGCSE Biology.

Activities

- Begin by asking students what their cells need for aerobic respiration and what they produce during aerobic respiration. Then pose the problem that cells need oxygen but they are deep inside your body, and they produce lots of carbon dioxide which is poisonous – how do they get the oxygen they need and get rid of the poisonous carbon dioxide before it kills them?
- Look at pages 96 and 97 with your students. If possible, have a large version of Fig 5.5 A on the board. Look at the diagram and read through the text about each of the areas of the gas exchange system. Explain to your students how the structure of each area is closely related to its function in the body.
- Students answer Q1. Then they **EITHER** answer Q2 **OR** do **Worksheet 5.5.1** where they label the diagram of the human respiratory system but do not have to draw it out themselves. This makes sure that they have an accurate diagram – they can also colour it in if they have time.
- Students complete **Worksheet 5.5.2**.
- Once students have completed the first two worksheets, they should answer Q3. Then give them **Worksheet 5.5.3**. Here students answer the questions and then look to find the correct answer in the wordsearch. This reinforces their knowledge and understanding of key biological terms and structures. They should begin the worksheet without reference to their books, to see how many of the terms they have understood and remembered, but then should look up anything they cannot answer.

5.5 Student Book answers

1. The exchange of gases in the lungs with oxygen from the air in the lungs moving into the blood and carbon dioxide moving from the blood into the air in the lungs.
2. Mouth and nose: make air warm and moist and remove microorganisms

   Trachea: tube though which air moves into and out of the lungs

   Bronchus/bronchi: tubes carrying air from the trachea to the lungs and from the lungs to the trachea

   Lung: organ where gas exchange takes place

   Alveolus/alveoli: air sac(s) where gas exchange takes place in the lungs

   Pleura: slippery membranes which allow lungs to move during breathing

   Diaphragm: muscle sheet dividing the body and involved in breathing.
3. The cells lining the bronchi are a good example of specialised cells. They make mucus which traps dust and dirt from the air. It also traps harmful microorganisms which might infect the lungs. These specialised cells are also covered in cilia, which are hair-like structures that move the mucus, dirt and microorganisms away from the lungs.

5.6 Breathing

Student Book pages 98–99

kerboodle

Online resources

- Worksheet 5.6.1: Resting breathing
- Worksheet 5.6.2: Breathing: diagrams and models
- Worksheet 5.6.3: Extension: Analysing resting breathing rates
- Interactive test

Prior learning

- Use scientific names and position of some of the major organs in the body
- Describe the main functions of some of the major organs and explain how they are essential

Objectives

- Describe the action of the diaphragm and intercostal muscles in breathing
- Understand the difference between breathing and respiration
- Describe the strengths and limitations of a model

Overview

In this lesson, students learn the difference between breathing, aerobic respiration and gas exchange. Three common misconceptions are:

- We breathe in oxygen and breathe out carbon dioxide.
- The lungs suck air in and squeeze air out when we breathe.
- The lungs are like a pair of balloons.

Taking the time to get these concepts clearly differentiated and fully understood in Stage 8 helps your students to succeed in CLSS and gives them the ideal foundation for their IGCSE Biology studies. This lesson also gives you the opportunity to revisit TWS ideas on the use and evaluation of models.

Activities

- Begin by asking students for their ideas about what breathing is, why they breathe and what happens when they breathe. Note down their ideas.
- Give students **Worksheet 5.6.1**. Do this as a class exercise, with you controlling the timing. Each student should end up with their own mean resting breathing rate.
- Read the paragraphs on page 98 headed 'Why do we breathe?' and spend some time looking at the pie charts showing the composition of inhaled and exhaled air with students. Emphasise that both the air we breathe in and the air we breathe out are a mixture of gases. It is simply the proportions of the gases that change. Students could use the data from the pie charts to show the same information as bar charts to practise data manipulation. Students complete Q1.
- Now consider breathing. Look back at students' suggestions on breathing. Make sure students know that breathing is simply moving air in and out of the lungs. Emphasise that the lungs are completely passive. They cannot inflate themselves or deflate. Also emphasise that the lungs are not balloons – that they are made up of millions of alveoli – tiny air sacs.
- Students can feel their own rib cages rise and fall as they breathe in and out. Talk through the stages of inhaling and exhaling and let students relate the changes to the movements of their own rib cages as they breathe. Students make notes and draw diagrams.
- If possible, demonstrate the bell jar model of the lungs shown on page 99. If not, simply look at Fig 5.6 C. Point out that it allows them to see the effect of the movements of the diaphragm. Ask students for advantages and limitations of this model. Students answer Q2 from page 99.
- Students complete **Worksheet 5.6.2**. They label the different diagrams/models to reinforce their knowledge of the structures of the respiratory system and how it works. They explore the strengths and weakness of the different diagrams and evaluate the strengths and weaknesses of the bell jar model of the chest and lungs.

Extension

Worksheet 5.6.3.

Homework

Worksheet 5.6.3 or Workbook page 40.

Key words

inhale, exhale, intercostal muscles

5.6 Student Book answers

1. a. Aerobic respiration uses oxygen to break down glucose into carbon dioxide and water, releasing energy in a controlled way to be used by the cells.

   Breathing moves air into and out of the lungs, bringing oxygen into the body for respiration and removing waste carbon dioxide.

   b. Oxygen content of inhaled air: 21% oxygen

   Oxygen content of exhaled air: 16%

   21 − 16 = 5%

2. a. Muscles between your ribs contract pulling your ribcage up and out → muscles of the diaphragm contract pulling it down and flat → volume of chest increases → pressure inside chest decreases → air forced into the lungs.

   b. Muscles between your ribs relax so your ribcage moves down and in → muscles of the diaphragm relax so it moves up and dome shaped → volume of chest decreases → pressure inside chest increases → air forced out of the lungs.

3. a. In the bell jar model, the balloons represent the lungs, the rubber sheet is the diaphragm and the bell jar is the chest.

   Breathing in: The rubber sheet is pulled down. The volume inside the bell jar increases, the pressure decreases and air is forced into the balloons so they inflate.

   Breathing out: The rubber sheet is pushed up, the volume inside the jar decreases, the pressure increases and air is forced out of the balloons.

   b. e.g. The lungs are not like balloons – they are like sponges. Each balloon is more like a single alveolus.

   The walls of the jar cannot move – the ribs are moved up and down by the muscles changing the volume of the chest.

   The diaphragm at rest is a domed structure – the rubber sheet is flat so the changes in volume are not as clear.

   The diaphragm is made of muscle which contracts – it does not have to be pulled down or pushed up.

   Any other sensible point.

5.7 The effect of exercise on the breathing rate

TWS

Student Book pages 100–101

Objectives

- Predict the outcome of an enquiry based on your scientific knowledge
- Describe trends and patterns in results
- Make conclusions by interpreting results and explain the limitations of the results

Overview

This lesson helps students to develop their skills in Thinking and working scientifically.

Students are given the results of a class investigation into the effects of exercise on breathing rates. This picks up on the exercise they did in lesson 5.6 calculating their own mean resting breathing rate, so they bring practical experience to the process. This gives you the opportunity to present your students with a considerable amount of data to analyse. There is a worksheet which enables students to carry out a variation of this investigation themselves.

Activities

There are several ways to approach this lesson:

- Work through pages 100–101 with your students. Discuss different aspects of the practical described, encouraging students to predict the outcome of

kerboodle

Online resources

- Worksheet 5.7.1 The effect of exercise on the breathing rate
- Worksheet 5.7.2 The effect of exercise on the breathing rate – questions 1–5
- Interactive test

Extension

If doing the practical, complete questions 1–5 from the Student Book or **Worksheet 5.7.2**.

Homework

Either **Worksheet 5.7.2** or workbook page 41.

the investigation based on their knowledge, look for trends in the results and make conclusions. They should also discuss limitations and answer questions 1–5.

- Work through pages 100–101 with your students. Discuss different aspects of the practical described, encouraging students to predict the outcome of the investigation based on their knowledge, look for trends in the results and make conclusions. Then give students **Worksheet 5.7.1** and ask them to compare the method described in the Student Book with the method described on the worksheet, evaluate them and decide which is the better method. Students then carry out the investigation to observe the effect of exercise on their breathing rate.

SAFETY WARNING: Make sure that students who do not take part in PE lessons, for health reasons, do not do this practical – perhaps suggest they take charge of the timing instead. Check that students who usually use an asthma inhaler before exercising use their inhalers before taking part in this practical.

Students write up their investigation and results. You can combine the whole class results in a table similar to Table 2 on page 101 and students can use their own data for the calculations.

- Tell students that they will be investigating the effect of exercise on their breathing rates. Ask them to predict what this will be using their knowledge of breathing and respiration. Give students **Worksheet 5.7.1** and ask them to carry out the practical. Apply the same safety precautions as described above.
- Once students have carried out the practical themselves, work through pages 100–101 of the Student Book with them. Discuss different aspects of the practical described, comparing it with the practical they have done themselves, evaluating the methods, and discussing the advantages and limitations of both. Students then answer questions 1–5, completing them from **Worksheet 5.7.2** as homework if necessary.

5.7 Student Book answers

1. **a.** Any movement uses energy and so increases the need for oxygen. This is likely to affect the breathing rate so it's important to be as still as possible before measuring the resting rate.
   b. They are investigating the effect of exercise on the resting breathing rate. It is important that they sit still after each exercise session to allow their breathing rate to return to its resting level or the results will not be valid.
2. Repeat measurements make the data more reliable – any single measurement may be anomalous.
3. **a.** Mean resting breathing rate per minute for the class: 280 ÷ 20 = 14 breaths/min; mean breathing rate per minute after exercise for the class: 380 ÷ 20 = 19 breaths/min.
   b Simple bar graph of the two mean figures calculated in **a** – check for labelled axes and labelled bars.
   c Individual students may make errors in counting/measuring time, etc. – but the more results that are combined, the less any individual errors will affect the pattern.
4. Resting breathing is affected by age, size, etc., so there will be variety in the class. It supplies the oxygen needed by the cells of the body at rest. When you exercise, the muscles need oxygen for respiration to provide the energy to contract – so the breathing rate must increase to supply the oxygen to be picked up by the haemoglobin in the red blood cells and carried to the cells of the body, especially the muscle cells. More carbon dioxide is produced as well, and increased breathing allows the body to remove this carbon dioxide through the lungs.
5. Limitations include: different students using different stopwatches – may not all measure the same time interval. Some students of different ages or sizes; some students exercise harder than others; some students may have a cold or something affecting their body; some students may miscount a breath or two – any sensible suggestions.

5.8 The structure of the alveoli

Student Book pages 102–103

kerboodle

Online resources

- Worksheet 5.8.1 The structure of an alveolus related to its function
- Worksheet 5.8.2 Gas exchange in the alveoli
- Interactive test

Extension

Ask students to evaluate the different sponges shown in Fig 5.8 A and 5.8 D as models of healthy lungs and lungs affected by COPD.

Homework

Workbook page 42

Key words

capillaries

Objectives

- Describe how the structure of the human respiratory system is related to gas exchange
- Describe the diffusion of oxygen and carbon dioxide between the blood and the lungs

Overview

In this lesson, students focus on the process of gas exchange in the alveoli. You have built up their knowledge and understanding of diffusion, the structure of the respiratory system, the process of aerobic respiration and the mechanics of breathing. You have also thought about the use of scientific models. Now you bring all of these elements together to enable students to understand the key process in the respiratory system – gas exchange in the alveoli. Emphasise the relationship between structure and function, and the importance of a large surface area for diffusion – both key principles that students will need to grasp to do well in IGCSE Biology.

Activities

- Ask students to describe diffusion, aerobic respiration and breathing.
- Explain to students that they will now find out how oxygen gets from the lungs to the cells, and how waste carbon dioxide is transported from the cells and removed into the air. This is the function of the alveoli.
- Look at the first paragraph on page 102 with your students, including Fig 5.8 A, and discuss the structure of the lungs. If possible, bring in a kitchen or bath sponge cut open to show the many tiny air spaces. Ask students to consider the strengths and weaknesses of this model of the lung tissue.
- Read through the section on 'Gas exchange and the structure of the alveoli', pages 102–103. Look at the diagrams (Fig 5.8 B) with students, and the micrograph in Fig 5.8 C. Talk about the relationships between the structure of the alveoli and their function. You could peel a large potato/yam and several small potatoes/yams to show students that the surface area (amount of peel) is much greater from several small 'spheres' than from one large one. This is a clear, physical demonstration of this important principle.
- Give students **Worksheet 5.8.1**. Ask students to add labels and notes to this diagram to show how the structure of each alveolus is adapted to its function of gas exchange. They should use their Student Books to support this activity. Alternatively, students answer questions 1–3.
- Give students **Worksheet 5.8.2** to complete. This demands a good understanding of the sequence of events during gas exchange in the alveoli and how it links to events in the cells and breathing air in and out of the lungs.
- Read through the final paragraphs on page 103 under 'COPD and the alveoli', and look at Fig 5.8 D. **Discuss this with sensitivity, as, with all health issues, some students will have family experience of these problems.** The key idea is that as the alveoli break down, there is less surface area for gas exchange and so the cells are short of oxygen and suffer a build up of carbon dioxide. Students answer Q4.

5.8 Student Book answers

1. One of the tiny air sacs which make up the structure of the lungs.
2. See Fig 5.8 B – the right hand image of the single alveolus. Correctly labelled alveolus, thin walls, red blood cells, oxygenated blood, deoxygenated blood, air in and air out, and arrows showing movement of carbon dioxide from the blood to the air, and oxygen from the air to the blood.
3. Function: gas exchange – oxygen from air to blood, carbon dioxide from blood to air. How structure relates to function: rich blood supply – diffusion of gases and maintains steep concentration gradients. Thin walls of alveoli and blood vessels: allow easy diffusion of gases; large surface area – lots of space for gas diffusion to take place.
4. If the structure of the alveoli break down, then the air sacs get bigger. Bigger air sacs mean smaller surface area – so less gas exchange takes place. A person who is short of oxygen/has a build-up of carbon dioxide feels breathless when they do anything that needs more oxygen.

5.9 Asthma

SIC

Student Book pages 104–105

kerboodle

Online resource

- Worksheet 5.9.1 Asthma awareness poster
- Interactive test

Prior learning

- How the structure of the human respiratory system is related to gas exchange

Objectives

- Discuss how specific knowledge is developed and used over time
- Describe how science is applied across societies

Overview

This lesson helps students to develop their understanding of Science in context. It gives you the opportunity to explore these ideas by looking at how our ideas on asthma and asthma treatment have developed over time and across societies. Asthma is a common condition affecting the respiratory system, especially in children, which is known about and treated around the world. Understanding asthma reinforces knowledge and understanding of the human respiratory system and how it functions. Considering how our knowledge of the condition has developed over time and the treatments developed enables students to see the relevance of what they are learning. **As always when tackling health issues, this needs sensitive handling. Most classes will have at least one student with asthma or with family members affected by asthma. Check with your students before this lesson. You may have students with asthma who prefer others not to know. Or you may have students with asthma who would be happy to tell their classmates what an attack feels like, or how they use their inhalers.**

Activities

- Begin by introducing the concept of asthma to your class. Look up the names of well-known local sportsmen and women who are affected by asthma but who have achieved sporting success.
- Read the first section of page 104 'What is asthma?'. Discuss the problems of developing treatments for a condition if you don't understand how it affects the body. Ask students to consider the difference it makes now that we know that the lining of the tubes of the respiratory tract swells and the muscles contract and narrow the airways during an asthma attack.
- Ask students for suggestions about how the symptoms of asthma are linked to the changes in the airways.
- Students read page 105 about the different types of asthma treatment available and the challenge of getting drugs to the airways where they are needed. Talk through the ideas, emphasising how scientific knowledge has developed over time and the difference that better medicines have made to people and societies around the world.
- Students answer questions 1–4.
- Give students **Worksheet 5.9.1**. They are asked to design a poster to inform other students about asthma. They could make this full size for display in the classroom. Encourage them to include as much clear biology in their posters as possible. Students can complete this for homework.

Extension

Carry out some online research into the history of the scientific understanding of asthma and the development of the different treatments. Then write a report/develop a timeline to inform others.

Homework

Complete poster or Workbook page 43.

Key words

inhaler, environmental triggers

5.9 Student Book answers

1. The muscles around the airways/bronchi/tubes to the lungs contract, narrowing the tubes and making it hard to move air into or out of the lungs. At the same time, the lining of the bronchi/tubes swells and produces a lot of extra mucus. This also makes the tubes narrower and so makes it harder to move air in and out. The person affected has a tight chest, wheezes, is short of breath, etc.
2. **a.** Any 3 from: pollen, pet hair, smoking, mould/spores, infections, food, medicines, dust mites, pollution, exercise, stress, cold air.
   b. 8 or 9
3. **a.** Relievers are used when someone has an asthma attack to make them feel better fast. Relievers relax the muscles around the bronchi. This opens up the airways quickly, making it easy to breathe again.
   Preventers reduce inflammation and so reduce the sensitivity of the airways, making asthma attacks less likely.
   b. After scientists understood the importance of muscles contracting in making the airways narrow in asthma, they made medicines to relax the muscles and relieve the symptoms. Scientists did not understand the importance of inflammation in asthma for another 100 years. They could not make medicines to reduce the sensitivity of the airways until they understood its cause.
4. **a.** Getting them into the right place in the airways.
   b. The development of inhalers which deliver the medicine into the respiratory system.

5.10 Transport in the blood

Student Book pages 106–107

kerboodle

Online resources

- Worksheet 5.10.1 Analysing blood
- Worksheet 5.10.2 Blood flow and the dangers of frostbite
- Worksheet 5.10.3 Extension: What can the blood tell us?
- Interactive test

Extension

Give students **Worksheet 5.10.3**. Students cut out the blood data cards and then look at the blood profiles of three different patients to diagnose their diseases. (Patient

Objective

- Describe the components of the blood and their functions

Overview

In this lesson, you introduce your students to blood as a complex liquid that carries out many functions in the human body. It is an excellent opportunity for helping your students to develop an holistic view of the body, with all the different systems linked by the blood. This lesson reinforces the idea that the blood carries oxygen from the lungs to the tissues, and carries carbon dioxide away from the cells back to the lungs to be removed from the body. The worksheets give students the opportunity to analyse data and draw conclusions.

Activities

- Ask students what they know about blood and list their answers.
- Read through the first half of page 106 with students. The key ideas are: 1) humans are large, multicellular organisms and so need a transport system to carry materials to and from the cells; 2) the blood is the main transport system; 3) the blood is made up of a number of different components. Have a micrograph slide of blood projected for reference.
- Students look at Fig 5.10 B and use the data to draw a pie chart to show the proportions of the different components of the blood. Ask them to annotate each section of the pie chart with the functions of that part of the blood.
- Remind students of the different materials that need transporting around the body in the blood – but also emphasise other functions of the blood such as defence against disease and clotting. Read with students the four sections under the heading 'The structure and functions of the blood'. Take time after reading about each blood component to explain how it works and answer any questions the students have, before they make brief notes.

1 = haemophilia (low platelet count); patient 2 = infectious disease e.g. flu (high white cell count); patient 3 = anaemia (low red cell count).

Homework

Workbook page 44

Key words

red blood cells, white blood cells, plasma, platelets, antibodies

- Students answer questions 1 and 2 on page 107.
- Give students **Worksheet 5.10.1**. They look at changes in the composition of the blood as it passes through different organs and are asked to describe and explain the patterns that they see.
- Now give your students **Worksheet 5.10.2**. Ask them to complete the activity, which pulls together the importance of the blood in supplying the food and oxygen needed by the cells and removing waste products.
- Finally, ask students to answer Q3 on page 107, thinking about the major functions of the blood and its importance in the body.

5.10 Student Book answers

1. Order of column 1 does not matter

Component	% volume	Function
Plasma	55	Transports the blood cells and platelets, transports dissolved substances including nutrients, carbon dioxide and urea around the body.
Red blood cells	44	Transport oxygen to the body cells.
White blood cells	1	Protect against pathogens that cause disease.
Platelets	Less than 1	Makes the blood clot when you are injured so you don't bleed to death. Help to produce scabs which protect the tissue and allow it to repair without infection.

2.

Substance	Part of the blood it is transported in	From	To
Oxygen	Red blood cells	Alveoli/lungs	Cells of the body
Carbon dioxide	Plasma	Cells of the body	Alveoli/lungs
Digested food	Plasma	Digestive system	Cells of the body
Urea	Plasma	Liver	Kidneys

3. Any four from:

- Blood transports the oxygen needed for respiration to all the cells of the body.
- Blood transports waste toxic carbon dioxide made in respiration away from the cells to the lungs to be removed from the body.
- Blood carries dissolved nutrients from the digestive system to the cells where they are needed for respiration, growth, etc.
- Blood removes the toxic urea made in the liver and transports it to the kidneys where it is removed from the body.
- Blood carries white blood cells to areas where pathogens have got into the body to prevent infectious diseases.
- Platelets make the blood clot if you have a cut so you don't lose a lot of blood.
- Platelets help your body make scabs which protect wounds and let the tissue repair itself.
- Any other sensible point.

Extension 5.11 Coping with extremes

Student Book pages 108–109

kerboodle

Online resource

- Worksheet 5.11.1
- Interactive test

Prior learning

- Different animals are found in different habitats and are suited to the environment they are found in

Homework

Workbook page 45

Key words

myoglobin

Objectives

- Describe the components of the blood and their functions
- Describe how the structure of the human respiratory system is related to its function of gas exchange

Overview

In this extension lesson, your students apply what they have learned about the structure and function of the human respiratory system to different extreme situations. The examples given in the Student Book include the challenges to people living at altitude and to large whales. Students have data to analyse and new terminology to cope with. This lesson moves them towards their IGCSE Biology course, where they have to recognise adaptive features and explain how they provide organisms with a survival and reproductive advantage.

Activities

- Begin by asking students to list some of the most extreme conditions that people live under. Ask them to decide which environment is hardest for the respiratory system. There is no correct answer to this, but in the Student Book they will look at one very tough environment – high altitude.
- Read page 108 with students and look at the graph in Fig 5.11 B. The main points are: 1) understanding why the reduced air pressure at altitude makes it hard to get enough oxygen from the air; and 2) the fact that the problems are different depending on whether you are born at altitude or simply go up a mountain. Students complete questions 1 and 2.
- Read through page 109 'The ocean deeps' with students, to help them with the terminology. This content looks at adaptations of whales as an example of a diving mammal. Students are introduced to some of the adaptations of the physiology of a whale to diving – giving them advanced terminology but describing aspects of biology which gives students the opportunity to apply what they have learned. Give students **Worksheet 5.11.1** and ask them to annotate the diagram of the whale and colour it to help them to remember the key points. If there is time, students can research more facts about humpback whales.
- Have a brief plenary session and ask students to apply what they have learned to complete Q3.

5.11 Student Book answers

1. Until 4000 m, % oxygen saturation of the blood stays relatively high; well over 90%; from 4000 m onwards the % oxygen saturation falls rapidly down to 85% at 4000 m [and into the 60s above 7000 m]. Lack of oxygen means you can't respire aerobically and so reduces the energy available. This will make you feel tired and unwell.
2. **a.** Gives a bigger volume inside the chest when breathing in, so the pressure can be lower so more air moves into the lungs when breathing, in spite of low external air pressure. More air = more gas exchange in the lungs.
   b. More airways mean more air is brought into the alveoli and more alveoli mean a larger surface area is available for gas exchange.
   c. More capillaries mean more blood flowing around the alveoli. This maintains a steep concentration gradient so more oxygen diffuses into the blood and more carbon dioxide diffuses out.
   d. Red blood cells take up oxygen in the capillaries of the lungs and carry it around the body, releasing it to the tissues and cells that need it for cellular respiration. Many extra red blood cells mean more oxygen carried around the body in spite of the low air pressure.
3. Most athletes are not born at altitude so, when they go to altitude training, their bodies must adapt to having less oxygen available in the air. Main adaptation = make more red blood cells to carry more oxygen. When they return to sea level to compete they have lots of extra red blood cells, which carry more oxygen even in normal air pressures and this allows them to perform better.

5.12 Review answers

Student Book, pages 110–111

Student Book answers

1. a. Aerobic respiration/cellular respiration **[1]**

b. glucose + oxygen → carbon dioxide + water + energy **[2]**

c. i. Microscopic structures in cells where aerobic respiration takes place.

ii. The cell with few mitochondria is not very active and doesn't need much energy; the cell with many mitochondria is a very active cell requiring lots of energy; the cell containing chloroplasts and mitochondria is a plant cell; aerobic respiration takes place in both animal and plant cells. **[4]**

2. a. D **b.** C **c.** F **d.** A **e.** E **f.** B **[6]**

3. a. alveolus/alveoli **b.** bronchus **c.** diaphragm **[1,1,1]**

4. a. Muscles between your ribs contract pulling your ribcage up and out → muscles of the diaphragm contract pulling it down and flat → volume of chest increases → pressure inside chest decreases → air forced into the lungs **[5]**

b. Muscles between your ribs relax so your ribcage moves down and in → muscles of the diaphragm relax so it moves up and dome shaped → volume of chest decreases → pressure inside chest increases → air forced out of the lungs **[5]**

5. a. Trachea **b.** Diaphragm **[1,1]**

c. When the rubber sheet is pulled down the balloons inflate. Pulling the rubber sheet down increases the volume of the model chest. This lowers the pressure inside the jar, so air is forced into the balloons by the pressure of the air outside the jar.

d. Any two from:

In the real chest, the ribs move up and out as a result of muscles contracting. The walls of the bell jar are fixed. In the real chest, the diaphragm is domed up into the chest when it is relaxed and flattens when you breathe in. The muscles of the diaphragm contract to pull the diaphragm down and flat – it isn't done by hand. The lungs are not like balloons. They are sponge-like, made up of millions of tiny air sacs which each individually act a bit like these balloons. Any other well made point. **[4]**

6. a. X = alveolus, Y = blood capillary **b.** Gas A is oxygen, gas B is carbon dioxide. **[2,2]**

c. Diffusion is the net/overall movement of particles from an area where there are lots of them (a high concentration) to an area where there are fewer of them (a lower concentration). **[3]**

d. There is only a short distance for the gases to travel so diffusion into or out of the blood happens easily. **[1]**

e. e.g. they have a large surface area – big area for gases to be exchanged; there is a rich blood supply bringing carbon dioxide to the lungs and picking up oxygen to take to the cells; there are steep concentration gradients between the blood and the air in the alveoli so diffusion happens as quicky as possible. Any other sensible points. **[3]**

7. a. B **b.** A **c.** C **[1,1,1]**

8. a. i. Platelets **ii.** White blood cells **iii.** Red blood cells **b.** Plasma **c.** 55% **[3,1,1]**

d. e.g. digested food, e.g. glucose; waste products/ urea; waste products/carbon dioxide; hormones; antibodies. **[3]**

9.

Gas	Inhaled air	Exhaled air
nitrogen	79%	79%
oxygen	21%	16%
carbon dioxide	0.04%	4%

The air breathed in/inhaled is relatively high in oxygen and low in carbon dioxide. In the lungs, gases are exchanged between the air in the alveoli and the blood. Oxygen moves into the blood by diffusion and is carried to the cells for aerobic respiration. Carbon dioxide, the waste product of respiration, is carried in the blood to the lungs. It diffuses into the air in the alveoli. As a result, the air breathed out/exhaled is lower in oxygen than the air breathed in, but it is higher in the waste product carbon dioxide. **[8]**

10. a. Use data on table. Correct axes. Show different readings for each individual. **[8]**

b. E is least fit – highest resting breathing and biggest increase in breathing rate after exercise. D is fittest – lowest resting breathing rate and lowest increase in breathing rate after exercise. **[2,2]**

c. Resting = 14; Exercise = 20. Use data in table. Correct axes **[2,2,3]**

d. Individual mean – enables you to see differences between individuals/identify most or least fit.
Group mean – increases reliability, enables you to see main trends **[2]**

6.1 The food we eat

Student Book pages 112–113

kerboodle

Online resources

- Worksheet 6.1.1 Testing food for protein
- Worksheet 6.1.2 Classifying food
- Interactive test

Prior learning

- Know that an adequate, varied diet is needed to keep healthy

Extension

Students investigate the main sources of protein in the food in their own culture and research the main protein sources in other cultures and then write a report on protein sources around the world.

Homework

Workbook page 46

Key words

diet, carbohydrates, protein, amino acids, lipids

Objectives

- Identify the constituents of a balanced diet for humans
- Understand that animals eat food to get energy and nutrients

Overview

In this lesson, you introduce students to the concept that animals, including people, eat food to get energy and nutrients. The important idea is that all the different foods we eat, in different countries and cultures, all contain the same basic food groups. There is a common misconception that the term diet applies to the food eaten when trying to lose weight. Make sure your students use the term correctly and understand that it means all of the food you take into your body.

Food tests provide plenty of opportunity for practical work by students or demonstrations by the teacher.

Activities

- Ask each student for their favourite food – try to get everyone to contribute.
- Ask students what they think when they hear the word 'diet'. Discuss the various answers you may get and read the first two paragraphs on page 112. Reinforce strongly that your diet is simply the food you eat every day.
- Question students about the different types of compounds that make up their food – have they heard of carbohydrates, proteins, etc?
- Read through the rest of pages 112 and 113 with your students. Stop at the end of each category and have a Q&A session in which you ask students to give examples of foods rich in the different food groups, and their ideas about what they are needed for, to check that they understand the content. If possible, have examples of foods rich in the different categories with you to make the point. Students complete questions 1 and 2 on page 113.
- Introduce your students to food tests used to identify the presence of different food groups and give them **Worksheet 6.1.1**. This gives two processes, using sodium hydroxide solution and copper sulfate solution OR using ready mixed Biuret solution. Ideally, your students should carry out these investigations themselves, developing practical skills and learning about food groups at the same time. If this is not possible, carry out the investigations as a demonstration. If that is not possible, read through the worksheet with students and find a video online which demonstrates the protein test.
- Give students **Worksheet 6.1.2**. They will use this diagram in this lesson and the next, filling in examples of different types of food on the Venn diagram. They add the results of their protein test investigation into foods containing protein, and also include their answers to Q3 on page 113.
- Summarise the learning so far and ask students to be very aware of the food they eat before the next lesson – maybe even note it down.

6.1 Student Book answers

1. The food you eat each day.
2. **a.** Carbohydrates, proteins, lipids/fats and oils, vitamins, minerals.
   b. Carbohydrates – energy; proteins – to build new cells/repair damaged cells; fats and oils – energy or energy store; vitamins – help chemical reactions take place in cells; minerals – to build different substances in the body.
3. Depends on what students ate – look for correct identification of foods containing the five main groups listed above.

6.2 Food and energy

Student Book pages 114–115

kerboodle

Online resources

- Worksheet 6.2.1 Testing food for starch
- Worksheet 6.2.2 Testing food for glucose
- Worksheet 6.2.3 Testing food for lipids
- Interactive test

Prior learning

- Know that an adequate, varied diet is needed to keep healthy

Objectives

- Understand that animals eat food to obtain energy and nutrients
- Understand that animals use both carbohydrates and fats to store energy

Overview

This lesson continues directly from Unit 6.1. It looks at the foods which contain the most energy, and the ways in which the body stores the energy it needs. Link this work on energy back to Unit 5.2 on aerobic respiration. It also provides more opportunities for students to carry out practical work with worksheets covering food tests for starch, glucose and lipids. Throughout the lesson, they can add more examples into the Venn diagram they used in the last lesson from **Worksheet 6.1.2**.

Activities

- Begin by asking students for examples of foods in their diet since the last lesson – ask them about food they have eaten that is rich in carbohydrates, or protein, or lipids, for example.
- Remind students that cells need energy to do work – from muscle cells contracting to building the chemicals needed for the body to function. Read through page 114 with students. Begin with carbohydrates – simple sugars like glucose, and starches. As in Unit 6.1, it is helpful to have some examples of the different foods to show students.
- Give students **Worksheet 6.2.1** and give them the opportunity to do a preliminary iodine test on starch solution, glucose solution and water. Then allow students to test a few different foods for starch, e.g. bread, potato, yam or breadfruit, and a piece of protein-rich food that does not contain starch (for example, egg white). As an alternative, demonstrate these investigations – research shows that good teacher demonstrations with plenty of class interactions are a very effective learning tool.
- Give students **Worksheet 6.2.2**. Give them the opportunity to carry out the preliminary Benedict's test using glucose solution, starch solution and water. Then **EITHER** give students time to carry out the Benedict's test on different foods to see if they contain glucose **OR** demonstrate the Benedict's test on a number of different foods to see if they contain glucose.
- Read through the content on Fats and oils with students. Give them **Worksheet 6.2.3** and allow them the opportunity to carry out this simple test to see how filter paper turns translucent when rubbed with a fat or oil.
- Discuss the need to store energy – we are not constantly taking in and digesting food, but our cells always need energy. Read through the final section on page 115 with students. Emphasise the difference between the carbohydrate glycogen stores in the muscles and liver that give us a quick burst of energy when needed and the main energy store in our fat cells.
- Students answer questions 1–4a.

Extension

Students complete Q4b comparing glycogen and fat stores in the body.

Homework

Workbook page 47

Key words

sugars, glucose, starch, glycogen

6.2 Student Book answers

1. Carbohydrates and lipids/fats and oils.
2. Similarity: both are carbohydrates.
   Differences: any two from – sugars are small molecules, starches are big molecules made up of lots of sugars; sugars are absorbed into the blood quickly, starches have to be digested before they can be absorbed; sugars for quick release energy, starch for slow release energy.
3. Seeds and nuts often contain large stores of fats and oils to provide energy for the new plant that grows. Animals eat them because they are a very energy-rich food.
4. a. Because they are not eating all the time so they need energy stores to release energy when they are not taking in food.
   b.

Glycogen	**Fat**
Carbohydrate	Lipid
Short-term energy store	Long-term energy store
Stored in the muscles and liver	Stored in fat cells under the skin and around body organs
Provides rapid instant energy when blood sugar levels fall	Provide energy over time if not enough food is eaten to supply our energy needs

6.3 Measuring the energy in food – managing variables

TWS

Student Book pages 116–117

Objectives

- Understand how scientists measure the energy stored in food
- Decide what equipment is required to carry out an investigation or experiment and use it appropriately
- Consider which variables should be changed, controlled or measured when you plan an investigation

Overview

This lesson helps students to develop their skills in Thinking and working scientifically.

Students will learn how we measure the energy stored in food and the units used: kilojoules and kilocalories. These are the SI units and the traditional units, still widely referenced on food packaging. You will explain how to measure the energy stored in food using a very simple calorimeter. Engage your students in a discussion about the different variables involved and how they can be controlled. You could give them a version of this practical to do themselves, demonstrate the technique yourself or use only the Student Book to generate the same discussion. There are also demonstrations available online.

This is a valuable opportunity to raise the problem of controlling all the variables when working with biological systems. Living things vary – for example, you can measure the same mass of date and peanut – but each date and peanut will be different from the next as a result of genetic and environmental variation. This means that in biology some variables cannot be completely controlled. This is one way in which biology differs from physics and chemistry.

Activities

- Bring a selection of food packaging into the lesson and let students look at the energy values of the different food and the different units used. Ask for ideas about why this might be, then read the first section on page 116 with students.

kerboodle

Online resources

- Worksheet 6.3.1 Which food gives us the most energy? – managing variables
- Worksheet 6.3.2 Which food gives us the most energy? – results, conclusions and evaluation
- Worksheet 6.3.3 Extension: Calculating which food stores the most energy
- Interactive test

- Give students **Worksheet 6.3.1**. Look at this alongside the section 'How do we measure energy in food?' on page 116 and discuss the similarities and differences in the practical techniques with students.
- Now look at the section headed 'Making a fair comparison'. Ask different students to read out each comment and then have a brief class discussion on the value of the statement.
- If possible, students carry out and write up the practical on **Worksheet 6.3.1**. They can also be given **Worksheet 6.3.2** to help them with their write up.
- If students cannot do the practical themselves, **EITHER** carry out a demonstration **OR** use an online demonstration.
- Students complete questions 1 and 2 on page 117.

Extension

EITHER Give students **Worksheet 6.3.3**, which has Q3 from page 117 plus another exercise **OR** simply ask them to complete Q3 on page 117.

Homework

Each student to make a record of the food they eat for at least one day before the next science lesson – ideally they record this for three days – and bring their records to the next science lesson.

Workbook page 48

6.3 Student Book answers

1. Place a measured volume of water in a test tube/boiling tube → weigh a food sample → record the water temperature → set the food sample alight and use it to heat the water → record the maximum water temperature reached → weigh any food left → subtract the final mass from the initial mass to find out how much of the food burned → subtract the initial temperature from the maximum temperature to find the temperature rise → substitute data into this formula:

   temperature rise per gram (°C/g) = measured temperature rise (°C) ÷ mass of food used (g)

   → ***repeat twice more and find the mean → repeat the whole process with a second type of food so you can compare the amount of energy in the two foods.***

   Also acceptable if students say weigh the same mass of two different foods and use the same amount of each rather than weighing before and after burning.
2. Volume of the water, mass of the food sample, any other sensible point.

   Additional variables: e.g. control the energy loss to the environment, use oxygen rather than air to ensure same amount available for burning, any other sensible point.
3. Total mass of date burned = 30 – 5 = 25 g

   Temperature rise from heating with date = 25 – 20 = 5 °C

   Temperature rise per gram = 5/25 = 0.2 °C per gram

   Total mass of peanut burned = 1.2 – 0.2 = 1 g

   Temperature rise from heating with date = 20.4 – 20 = 0.4 °C

   Temperature rise per gram = 0.4/1 = 0.4 °C per gram

   Peanuts had more energy per gram as dates.

6.4 A balanced diet

Student Book pages 118–119

kerboodle

Online resources

- Worksheet 6.4.1 Eating a balanced diet
- Worksheet 6.4.2 A healthy food plate
- Interactive test

Prior learning

- Know that an adequate, varied diet is needed to keep healthy

Objectives

- Identify the constituents of a balanced diet
- Discuss how human development, growth and health can be affected by diet

Overview

Unit 6.4 focuses on the concept of a balanced diet in terms of a variety of food intake. A common misunderstanding is that the diet must be balanced every day. Emphasise that very few people have a perfectly balanced diet every day, but that over weeks or months the diet must be balanced to avoid health problems. You also introduce the importance of water and fibre in the diet. Neither are nutrients but both are needed for a healthy body.

Activities

- Ask students what is needed in a healthy diet – they should remember the main food groups they have met so far – carbohydrates, proteins, lipids (fats and oils), vitamins and minerals. Then look at Unit 6.4 in the Student Book and read the section on a balanced diet.
- Ask lots of questions to make sure that students understand the principle of a balanced diet and recognise that different people will need different balances of foods at different life stages, e.g. if they are growing fast, if they are pregnant, if they do a very physical job, if they are training for a sports event.
- Look at the food pyramid in Fig 6.4 B as an example of a balanced diet. Make the point very clearly that most people do not eat a balanced diet every day, but that over time the diet must be balanced to be healthy.
- Ask students to get out their record of the food they have eaten and give them **Worksheet 6.4.1**. If students have recorded their food over several days, give them more than one copy of the worksheet or suggest they redraw and extend the table so they can see how their diet balances over several days. Students analyse their diets and look at possible changes they might make. Remind them again that a balanced diet happens over time, not on a single day – **be sensitive to the circumstances of individual students, especially if you are doing this lesson during Ramadan or you have students affected by conditions such as diabetes in your class.**
- Students read through the section 'What else do we need?' to the end of the second bullet point. Question them on the importance of water and fibre in the diet.
- Give students **Worksheet 6.4.2**. Remind them of the food pyramid in Fig 6.4 B. Comment on the difficulty of educating people about the importance of a balanced diet. Ask them to design a 'Healthy food plate' – using the circle as a plate but also as a pie chart to show people what proportions of different types of food are needed to make a balanced meal. Their food plate must be bright and interesting and carry simple information for people to understand and use.
- Leave time at the end of the lesson to introduce the ways in which diet can affect human health (students can complete their food plate for homework). Read through the final paragraph, carefully considering the data in Fig 6.4 D. If time, students answer questions 1–3.

Extension

Students answer Q4 on page 119.

Homework

Students complete **Worksheet 6.4.2** or Workbook page 49.

Key words

balanced diet, fibre, cellulose

6.4 Student Book answers

1. A balanced diet contains a wide variety of foods that give you the energy you need to replace the energy used by your cells to carry out the functions of life, and the nutrients your body needs to grow and repair itself.
2. A balanced diet contains more carbohydrates, fruits and vegetables than anything else. It should contain fibre in the wholegrains and fruits and vegetables. Protein should be the next biggest constituent of the diet to replace cells, etc. There should only be a small percentage of fats and oils, and sweet or fatty treats.
3. In a cold country: needs more energy to keep warm and hence more high energy food. Loses less water as sweat to cool down, so less water in their balanced diet.

   In a hot country: needs less energy rich food, as no issues with keeping warm, but more water, as lose sweat to keep cool.
4. The risk of dying of colon cancer over 25 years drops from about 1.5% to about 0.5% with an increase in the fibre content of the diet from 20 g per day to 60 g per day. It falls in relation to an increase in fibre in the diet.

6.5 Diet, growth and development

Student Book pages 120–121

kerboodle

Online resource

- Worksheet 6.5.1 A balanced diet crossword
- Interactive test

Prior learning

- Know that an adequate, varied diet is needed to keep healthy

Objective

- Discuss how human development, growth and health can be affected by diet

Overview

In this lesson, you introduce students to the concept of deficiency diseases – as always when discussing health related issues, sensitivity is needed. Help your students to recognise that substances such as vitamins are needed in tiny amounts compared with carbohydrates, proteins and lipids, but they are nevertheless very important to the healthy working of the body. This spread focuses both on what vitamins and minerals are important for and the deficiency diseases that result if they are lacking in the diet. It is very impressive for students if you bring 2 chicken bones to the lesson – one normal one and one that you have stored covered in vinegar for around 5 days. The vinegar removes the calcium and so the bone is bendy – you can use it to show how important it is for bones to take up the calcium they need.

Activities

- Quiz students to find out what they know about vitamins and minerals with questions such as 'What is a vitamin?', 'What is a mineral?', 'Name a vitamin and tell us what it does in your body', 'What foods are rich in vitamins?' etc.
- Read through page 120 and the top of page 121 with students looking at the information on the role of vitamins in the body. Bring in some examples – citrus fruits, carrots or sweet potatoes, eggs. Look up stories about the discovery of vitamins and their importance to tell to students – for example, the use of citrus fruits to prevent scurvy on long sea voyages, the problems of rickets in countries where there is little sun, etc.
- Talk about importance of minerals such as calcium and iron in the diet. Link calcium to vitamin D and help students understand that vitamin D is needed for the bones and teeth to take up the calcium they need to stay strong. If you have prepared a chicken bone with the calcium removed, demonstrate it to students now – they will be amazed to see how flexible the bones are without calcium.

Extension

Once students complete **Worksheet 6.5.1**, they answer Q4 on page 121, on the role of certain vitamins and minerals in aerobic respiration.

Homework

Workbook page 50

Key words

deficiency disease, Vitamin A, night-blindness, Vitamin C, scurvy, Vitamin D, rickets, calcium, iron, anaemia

- Students work through questions 1–3 on page 121.
- Give students **Worksheet 6.5.1** after they have completed the questions. They should close their books and attempt to answer the questions and complete the crossword using their knowledge and understanding of vitamins, minerals and deficiency diseases. If they cannot complete the crossword, they should look up the answers in their Student Books – ask them to fill in any answers they have to look up in a different colour. This will emphasise what they have and have not learned and remembered.

6.5 Student Book answers

1. A disease caused by a lack or deficiency of a particular nutrient in the diet.

2.

Deficient substance	**Role in the body**	**Food containing this nutrient**	**Name and symptoms of deficiency disease**
Vitamin A	Supports healthy eyes	Egg yolks, fish liver oils, dairy products, carrots	Night blindness: poor vision in dim light
Vitamin C	Helps cells stick together/ aerobic respiration	Citrus fruits	Scurvy: gums bleed, skin peels, legs swell, cellular respiration stops
Vitamin D	Helps bones absorb calcium	Egg yolks, cream, fish liver oils	Rickets: soft, bendy bones, weak bones, crumbing teeth
Calcium	Makes bones and teeth strong	Dairy products	Rickets: soft, bendy bones, weak bones, crumbing teeth
Iron	Needed for haemoglobin in red blood cells	Red meat, egg yolks, liver, apricots, green leafy vegetables	Anaemia: not enough haemoglobin in red blood cells (rbcs), not enough rbcs = not enough oxygen – feel tired and weak.

3. Your body needs calcium to maintain strong, hard, healthy bones and teeth. It also needs vitamin D to absorb calcium from your digestive system. A diet low in calcium will cause rickets. But a diet high in calcium but low in vitamin D will still cause rickets because the body cannot absorb the calcium it needs.

4. Scurvy because it is a lack of vitamin C and vitamin C is needed for the reactions of aerobic respiration. Anaemia because it is a lack iron which is needed to make the haemoglobin which brings the oxygen needed for aerobic respiration to the cells. They are both very serious because aerobic respiration releases the energy needed for all the processes of life in your cells. If they cannot take place, you will eventually die.

6.6 Obesity and health

Student Book pages 122–123

kerboodle

Online resources

- Worksheet 6.6.1: Meeting our energy needs
- Worksheet 6.6.2: Extension: Eating behaviours
- Interactive test

Prior learning

- Know that some foods can be damaging to health e.g. very sweet and fatty foods

Objective

- Describe how lifestyle factors such as diet affect human health

Overview

This is another lesson that requires sensitivity – you may have students who are either underweight or obese in your class. No student should be made to feel uncomfortable during the lesson – but it is still important that students recognise the health problems linked to both conditions. This content is important for success at both CLSS and IGCSE Biology and understanding the links between diet and health is important for producing healthy and scientifically literate citizens of the future.

Activities

- Students list possible reasons why some people around the world do not have enough to eat. Put their answers together in a list on the board and then read the first two paragraphs on page 122 with your students. Discuss these reasons and recognise that, for most people, not having enough to eat is not a lifestyle choice – it is imposed on them. Remind students of the deficiency diseases they met in the previous lesson and explain that people who are starving are at risk of these deficiency diseases and many more.
- Explain that starvation is a problem for millions of people – but many more people around the world are affected by obesity than by being underweight. Obesity as a result of eating too much is a lifestyle choice for many people – but some people have other problems which mean they become obese very easily.
- List the main factors that increase the risk of obesity: eating more food than needed, eating food high in carbohydrates and fat, and lack of activity/exercise. Discuss them with your students. Aim to build understanding that obesity is a multifactorial problem and that societal changes have an impact on it as well as individual choices.
- Work through the BMI calculation with students. To avoid the chance of causing distress, it is better to do calculations with theoretical students – see Q4 later. Use Fig 6.6 B to help students understand the concept of risk, the increased risk of health issues with increasing obesity and that being underweight also increases the risk of disease.
- Students work through page 123 and use the data to make a table showing the increased health risks linked to obesity.
- Give students **Worksheet 6.6.1** to help them to understand how our energy needs change throughout different stages of life and depend on what we are doing. Once they finish the worksheet, they move on to answer questions 1–3.
- Finally, ask students to look at Q4 on page 123. All the students in the question have the same body mass. This exercise reinforces the concept that it is our level of obesity, not our absolute mass, that affects our risk of disease.

Extension

Give students **Worksheet 6.6.2**. This challenges them to look at how different behaviours affect how we eat, and to think of actions at both the societal level and the individual level that would help make more people healthy eaters.

Homework

Workbook page 51

Key words

obese, diabetes, high blood pressure, heart disease, heart attack, cancer, arthritis

6.6 Student Book answers

1. Obesity is carrying too much fat on our body and having a BMI over 30.
2. People are eating more, especially food high in carbohydrates and fats; people are exercising less as fewer people do physical jobs and more people drive rather than walk.
3. Any three from
   - Diabetes: blood sugar levels are hard to control. This causes tiredness and eventually blindness and even death.
   - High blood pressure: obesity is a common cause of high blood pressure. This stresses your heart and blood vessels and causes heart disease and strokes.
   - Heart disease: fatty material builds up in the blood vessels – you may have a heart attack.
   - Cancer: at least 13 different cancers are more common in overweight and obese people.
   - Arthritis: strain on the skeleton leads to wear and pain in the joints.
   - Deficiency diseases: fatty, sugary foods are often low in vitamins and minerals leading to deficiency diseases.
4. **a.** **i.** BMI = 70 ÷ (1.83 × 1.83) = 70 ÷ 3.35 = 20.9. Range 18–24.9.
   ii. BMI = 70 ÷ (1.50 × 1.50) = 70 ÷ 2.25 = 31.1. Range 30–39.9.
   iii. BMI = 70 ÷ (1.65 × 1.65) = 70 ÷ 2.72 = 25.7. Range 25–29.9.
   b. Jamil is in the healthy weight range so he is likely to be healthiest. Mo is overweight and Jasmin is obese – the higher your BMI, the more it is likely to affect your health.

6.7 Smoking and health

Student Book pages 124–125

kerboodle

Online resources

- Worksheet 6.7.1 The chemicals in cigarette smoke
- Worksheet 6.7.2 Smoking and lung damage – poster
- Worksheet 6.7.3 Smoking and lung damage – questions
- Worksheet 6.7.4 Extension: Changing their minds
- Interactive test

Objective

- Describe how lifestyle factors such as smoking affect human health

Overview

In this lesson, your students will discover how tobacco smoking affects health. This is a sensitive topic as many students may have family members who smoke or chew tobacco. The scientific evidence for the increased health risks from smoking are unequivocal and students may find this difficult to accept. Prepare the 'smoking machine' described in **Worksheet 6.7.1** before the lesson and use it as a demonstration – it is very striking. This lesson and the next also covers many of the objectives of Science in context as it looks at risk and the way scientific knowledge and human behaviour interact.

Activities

- An impactful way of starting this lesson is to ask students what they would feel if you told them to collect some cancer-causing leaves, set fire to them and breathe in the smoke.
- Read the first two sections on page 124 with your students, down to 'How does smoking make you ill?' Demonstrate the production of tar and acidic smoke using the smoking machine prepared earlier, and discuss the different compounds. Remind them that you are focusing on 3 out of about 4000 compounds in cigarette smoke. Then ask students to complete **Worksheet 6.7.1**.
- Read through the section 'How does smoking make you ill?' with your students. Emphasise that smoking does not mean you WILL get these diseases – it INCREASES THE RISK that you will be affected. Students answer questions 1–3 on page 125.

- Give students **EITHER Worksheet 6.7.2 OR** (if time is short) **Worksheet 6.7.3**. In both cases, students are given information about a smoker who has to use oxygen as a result of lung damage caused by smoking. Students use the information they are given along with information from the Student Book to design a poster warning of the issues of smoking-related lung damage or fill in a series of statements. If students design a poster, you can have a class vote to see which is thought to be the most effective.

Extension

Students use **Worksheet 6.7.4**. Their task is to produce a short presentation designed to prevent young people from starting to smoke. They need to demonstrate both understanding of the risks and the reasons young people might begin smoking.

Homework

Workbook page 52

Key words

nicotine, tar, carcinogen, carbon monoxide, lung cancers

6.7 Student Book answers

1.

Chemical	Damage caused
Tar	Coats the lungs making gas exchange difficult/makes the alveoli break down which makes gas exchange less efficient/contains carcinogens/cancer causing chemicals.
Nicotine	An addictive drug which makes it hard to give up smoking.
Carbon monoxide	Poisonous gas which joins to haemoglobin so there is less oxygen for the cells/less oxygen for developing baby in the womb.

2. a. Around 7 million.

b. It makes them feel good, tobacco smoke contains nicotine which is an addictive drug.

3. a. It contains many chemicals including tar and some of those chemicals are carcinogens – they affect the cells and cause cancer. These chemicals are taken into the delicate lungs and stay on the tissues for years.

b. If you smoke, most of the chemicals are drawn into the throat and particularly the lungs and stay there. The carcinogens affect the cells and cause cancer. If you chew tobacco, it contains the same chemicals but you keep them in your mouth, against the delicate tissues there. So the carcinogens dissolve in the saliva and affect the cells of the mouth, causing mouth cancers.

6.8 Building the evidence

TWS

Student Book pages 126–127

kerboodle

Online resources

- Worksheet 6.8.1 Evidence is not everything
- Worksheet 6.8.2 Extension: Smoking and lung cancer on a global and local scale
- Interactive test

Objectives

- Describe how lifestyle factors such as smoking affect human health
- Describe how scientific hypotheses can be supported by evidence from an enquiry

Overview

This lesson helps students to develop their skills in Thinking and working scientifically.

Students are presented with the story of the work of Sir Richard Doll who first made the link between cigarette smoking and an increased risk of developing lung cancer. They meet the concept of a peer-reviewed journal and consider how more evidence was collected to confirm the hypothesis of a causal link. This also gives you evidence-based answers to two common questions from students: what about people who get lung cancer but never smoked, and what about people who smoke heavily for many years and never develop any smoking-related health problems? The type of data analysis and understanding introduced on these pages are important skills for IGCSE Biology.

As always, when teaching health-related topics, this work should be handled with sensitivity.

Activities

- Begin by telling students how attitudes to smoking have varied over time – there are excellent quotes from different parts of the world at different times in history. Sometimes it was regarded as a horrible habit. More often it was regarded as something good for the health, sophisticated and enjoyable.
- Read the text on page 126 with students, finishing before 'More evidence'. When you cover the questions asked by the students, take time to discuss with your class the way people are often more influenced by their own experience than by scientific facts.
- Discuss the work of Sir Richard Doll and explain what is meant by a peer reviewed journal.
- Students read through the final paragraph on page 126 and answer Q1 on page 127.
- Students read through 'Changing evidence' on page 127 and then answer Q2.
- Ask students the question 'Why do people still smoke'? before they read the remaining text. Listen to their ideas and note them down on the board. Read through the final paragraph and bullet points and then give students **Worksheet 6.8.1**. Students look in more detail at the evidence in Fig 6.8 C and consider some of the reasons why the link between smoking and lung cancer was not easy to identify.
- Students answer Q3 on page 127, completing this for homework if necessary.
- Have a plenary session, to confirm that students have understood how evidence for the links between smoking and disease have built up over time, and that some people either develop lung cancer without smoking or smoke without developing the disease.

Extension

Worksheet 6.8.2

Homework

Complete Q3 on page 127 if not finished during class and/or Workbook page 53.

Key words

peer-reviewed, causal link

6.8 Student Book answers

1. Any one of the pieces of evidence shown in this spread which show a link between numbers of cigarettes smoked and deaths from lung cancer.
2. Hassan's grandfather has smoked heavily for a long time. In theory, this greatly increases his risk of developing lung cancer BUT: a) it only increases the risk, it does not mean he WILL get lung cancer; and b) some people have a genetic combination which protects their lungs so they are very unlikely to get lung cancer even if they do smoke.

   Up to 90% of lung cancers are linked to smoking but that means up to 10% of lung cancers are not linked to smoking. Samira's grandmother was one of this unfortunate 10%.
3. **a.** Two bar charts – students should title one bar chart Annual death rate per 100 000 people or have that label on the vertical axis and have Relative risk of developing lung cancer as the title or on the vertical axis of the other bar chart.

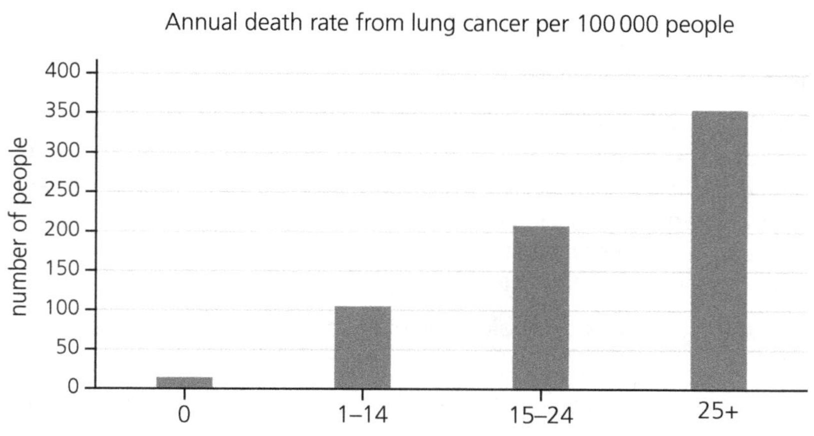

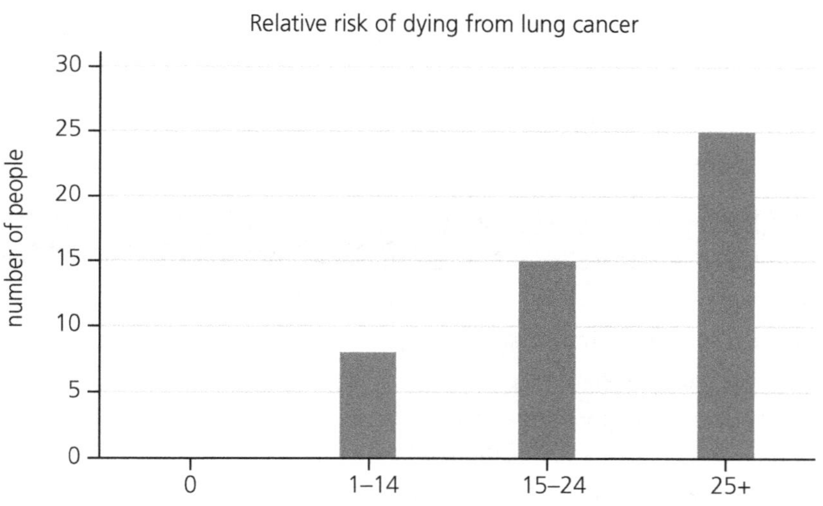

   b. In both sets of data, the trend is that the more cigarettes a person smokes, the higher their risk of dying from lung cancer. Any other sensible conclusions.

6.9 The human skeleton

Student Book pages 128–129

kerboodle

Online resource

- Worksheet 6.9.1 The human skeleton
- Interactive test

Prior learning

- Know that humans and some other animals have bony skeletons inside their bodies
- Know how skeletons grow as humans grow, support and protect the body

Extension

Students do research and discover how the basic design of the skeleton is different in birds, whales and snakes.

Homework

Workbook page 54

Key words

skeleton, vertebrae, joints, bone marrow

Objectives

- Identify the bones of the skeleton
- Describe the functions of the skeleton including the joints

Overview

The best way to approach this lesson is to build it around a skeleton – real, plastic or paper. If that is not available, have a large skeleton projected on the wall. Gather students round and talk them through the bones of the skeleton and the functions of the skeleton in the context of the one on display and their own bones. This is an excellent opportunity to bring in student's own ideas and experiences, e.g. of broken bones. Students cover the skeleton in some detail in Stage 4 – although they are unlikely to remember this in much detail, it is useful to reference that and get an idea of what they recall.

Activities

- Get students to sit in a semicircle around a model skeleton or a projected skeleton and stand by the skeleton yourself. Begin by asking students what they know about the skeleton and what it does. Be prepared for them to know quite a lot or very little. Make the most of this session – students are usually fascinated by the skeleton and this is potentially a very enjoyable and exciting lesson from which they learn a lot.
- Talk about the functions of the skeleton in the body as listed on pages 128 and 129.
- Go through the bones of the skeleton. Ask students to name as many as they can. Discuss the functions of different bones and relate their structures to their functions. Give students opportunities to share information and feel as many of their own bones as they can so that they can relate the skeleton to their own bodies.
- Give students **Worksheet 6.9.1** – they should label the diagram fully.
- Students answer questions 1–3 on page 129.

6.9 Student Book answers

1. Support, protection, movement, making blood cells.

2. a. Backbone (vertebrae)/femur/tibia/toe bones

b. Ribs, skull, backbone (vertebrae)

3. Long bones – compact bone on the outside – hard and strong.

Spongy bone inside – keeps them relatively light.

Bone marrow – makes red blood cells to carry oxygen around the body and white blood cells to protect against infection.

6.10 Muscles and movement

Student Book pages 130–131

kerboodle

Online resources

- Worksheet 6.10.1 Tissues in a joint: a chicken wing
- Worksheet 6.10.2 Observations on the tissues in a joint: a chicken wing
- Worksheet 6.10.3 Human muscles and tendons
- Worksheet 6.10.4 Extension: Mending damaged tissues
- Interactive test

Prior learning

- Know that animals with skeletons have muscles attached to the bones
- Know how a muscle has to contract (shorten) to make a bone move and that muscles act in pairs

Extension

Students carry out independent research using the internet to answer the questions in **Worksheet 6.10.4**.

Homework

Workbook page 55

Key words

hinge joints, ball-and-socket joints, fused joints, ligaments, cartilage, lubricates, tendons, contracts, antagonistic muscles, biceps, triceps

Objectives

- Identify different types of joints in the human body
- Explain how antagonistic muscles move the bones at a hinge joint

Overview

This lesson follows on from the previous work on the skeleton. Here you explain the importance of joints in movement and cover the main types of joints. You introduce students to the different tissues in a joint – a good opportunity to relate structure to function very clearly. A class dissection or demonstration of the tissues in a chicken wing reinforces these points strongly. Finally, discuss the role of antagonistic muscles in moving joints. Students often have many misconceptions about how muscles work. By clear explanations and activities using their own muscles you can ensure that students move on to IGCSE with a clear understanding of the basics of antagonistic muscles in the body.

Activities

- Ask students to visualise a solid skeleton – it would be good at protection, support and even making blood, but movement would be impossible. Use your model skeleton or projected skeleton to show students the main types of joints in the body and their importance for different types of movement. Students can move parts of their own bodies to see the different types of movement in hinge joints (fingers, elbows and knees) compared with ball-and-socket joints (shoulders and hips). Students read the descriptions of different joints on page 130 and draw and label a diagram of a hinge joint.
- Read through page 130 and the first paragraph of page 131 with students. Encourage them to link the properties of the tissues to their function in the joint.
- **EITHER** give students **Worksheet 6.10.1** and **Worksheet 6.10.2** where they carry out a dissection of a chicken wing and examine the tissues before answering the questions on worksheet 6.10.2.
- **OR** give students **Worksheet 6.10.2** and then carry out a demonstration of a chicken wing dissection yourself, pointing out the characteristics of the different tissues, before they answer the questions on the worksheet.
- **Take all safety precautions to avoid students getting raw chicken in their mouths.**
- Read through the remaining text on page 131 with students. Make clear that muscles can only pull, not push. Students work through **Worksheet 6.10.3**, completing the activities and labelling the diagrams.
- Students answer questions 1–4.

6.10 Student Book answers

1. **a.** Any two from: elbow, knee, ankle.
   b. Hip and shoulder.
   c. Hinge joint allows forward and backward movement, ball-and-socket joint allows all round movement.
2. Diagram based on the left hand diagram of Fig 6.10 A.
   Labels: Ligaments keep the joint in place; cartilage and fluid help the bones move smoothly in the joint.
3. **a.** Antagonistic muscles are pairs of muscles which work against each other to bend and straighten a joint.
   b. A muscle can only contract and pull in one direction. It cannot push. We need one set of muscles to pull the bone in one direction and another set of muscles to pull it back to its original place.
4. Biceps contract and shorten and triceps relax → lower arm holding the book is lifted up → biceps relaxes and triceps contracts and shortens → biceps returns to original length and lower arm carrying the book is lowered back down to the table.

6.11 Health and inequality

SIC

Student Book pages 132–133

kerboodle

Online resource

- Worksheet 6.11.1 Using scientific understanding to save the world
- Interactive test

Homework

Workbook page 56

Key words

inequality of access

Objective

- Evaluate issues which require scientific understanding

Overview

This lesson helps students to develop their understanding of Science in context. They look at inequality of access to food and education through the work of two students. This content helps your students to recognise that scientific understanding is required to tackle some of the biggest issues of our time. The evidence presented demonstrates that there is no need for people to have health issues as a result of too much or too little food – but equality of access to scientific understanding of the situation through education is crucial so that everyone has the opportunity to make informed decisions about their life.

Activities

- Read through the introductory section with students and ask: 'When scientists know so much about the impact of lifestyle choices on health, why are so many people still affected by diseases caused by their diet, or their habits such as smoking?'
- Students read through the work of Raj and Mariam and answer questions 1–3.
- Give students **Worksheet 6.11.1**. Students work in groups, although they could work individually, to develop a short news report about the importance of scientific understanding in enabling both people in power, and ordinary individuals, to evaluate problems and make healthy decisions. Students record their reports in the style of an audio or video news broadcast. If possible, enable students to watch each other's presentations and discuss which are the most effective at getting the message across.

6.11 Student Book answers

1. 45%–36% animal feed and 9% biofuels.

2. Any 3 valid reasons, e.g.

- Difficult to grow crops in some parts of the world, e.g. too hot, too dry. Solution: develop plants that grow in difficult conditions/improve farming methods.
- Only 55% of the crops grown are used to feed people directly – 36% is fed to animals to produce meat. Solution: people eat less meat, so more crops are available to feed people.
- 9% of crops grown are used to produce biofuels. Solution: stop using food which could be eaten to produce biofuels.
- Around 33% of all food produced is wasted or spoiled in storage. Solution: reduce food waste and improve food storage.
- Any other sensible suggestions.

3. Look for students referencing evidence and investigations, e.g. into different types of crops. Also look for reference to the role of education in both growing food and in understanding the importance of a balanced diet for healthy growth and development of children and for overall health. A good answer should have at least 6 well-made points.

6.12 Review answers

Student Book pages 134–135

Student Book answers

1. a. A balanced diet contains a variety of foods that supply the energy you need for your cells and the nutrients your body needs to grow and repair itself. **[2]**

   b. Carbohydrates provide both rapid energy and slow release energy sources, e.g. rice, bread (any example of carbohydrate rich food is acceptable). **[4]**

   c. Proteins are important to build new tissues, repair damaged tissues and replace old tissues, e.g. meat, fish, beans (any example of protein food is acceptable). **[4]**

   d. Lipids/fats and oils are important as an energy source and to produce insulating layer under the skin/protective layer around organs and cell membranes, e.g. butter, olive oil, cheese (any 2 examples of high fat food are acceptable). **[4]**

2. a. 50 g of fat. **[1]**

   b. Cheese contains 50% fat. Fat is very high energy food – it contains a lot of calories. When people want to lose body mass, they must reduce the amount of energy they take in. So reducing cheese would quickly cut a lot of energy from the diet. **[1]**

   c. i. 15 g more fat in meat than in fish. **[1]**

   ii. It is low in fat (relatively low in calories) and high in protein. **[2]**

3. a. Potato chips would supply most energy per gram as they have a much higher percentage of fat in them, and fat is a rich energy source. **[2]**

   b. If you record the temperature of a sample of water, then burn a food sample and use it to heat the water and measure the rise in temperature, this will give you a measure of the energy contained in the food. **[4]**

   c. Size of food sample (e.g. the same each time); the volume of the water heated; the distance of the flame from the tube; any other sensible points. **[2]**

   d. Not all the food burns away; some of the energy is lost as light and by heating the air. **[2]**

4. a. Lacking vitamin A. It is needed for healthy eyes and deficiency means you cannot see well in dim light. **[3]**

   b. Lacking iron. Iron is needed to make haemoglobin that carries oxygen to the cells for aerobic respiration which provides the cells with energy. With too little iron, a person makes fewer red blood cells so their skin is pale and they lack energy/feel tired/are breathless as they try to get more oxygen. **[4]**

   c. Lacking vitamin C. This is needed for cells to stick together. Lack causes scurvy where skin breaks up, gums bleed and teeth are lost. **[3]**

   d. i. Lack of calcium or lack of vitamin D. **[2]**

   ii. Calcium is found in milk and dairy produce. As the patient drinks lots of milk it is unlikely that the cause of her rickets is lack of calcium. Vitamin D comes from the diet but is also made in the skin when out in the sun. You need vitamin D to absorb calcium from the diet into the body. The patient stays indoors most of the time so doesn't get sun on her skin. This suggests the cause of her weak bones is lack of vitamin D not lack of calcium. **[5]**

5. a. BMI = Body Mass Index; calculated: BMI = mass/(height)^2 **[2]**

   b. BMI over 30 is obese. Obesity increases the risk of developing many different diseases. **[2]**

   c. Any three from:

   Diabetes: a lot of body fat makes blood sugar levels hard to control. This causes tiredness and eventually blindness and even death.

   High blood pressure: obesity is a common cause of high blood pressure. This stresses your heart and blood vessels and causes heart disease and strokes.

Heart disease: fatty material builds up in the blood vessels – you may have a heart attack.

Cancer: at least 13 different cancers are more common in overweight and obese people.

Arthritis: strain on the skeleton leads to wear and pain in the joints.

Deficiency diseases: fatty, sugary foods are often low in vitamins and minerals leading to deficiency diseases. **[6]**

6. a. The first is a ball-and-socket joint and the second is a hinge joint. **[2]**

b. Ligaments – hold the bones of the joint in place while allowing it to move; cartilage – cushions the ends of the bones and allows them to move smoothly over each other; fluid – keeps the cartilage wet and slippery so the bones move easily/no friction. **[6]**

c. A muscle can only contract and pull. It cannot push. So you need one muscle on one side of the joint to contract and bend the joint while the other muscle relaxes. This muscle then relaxes and the other contracts to move the bones in the other direction and straighten the joint. **[3]**

d. Antagonistic muscles **[1]**

7. a. Material in your food which cannot be digested e.g. cellulose/plant cell walls. **[2]**

b. USA and Uruguay. **[2]**

c. Title: Chart to show the percentage of high-fibre foods eaten in different countries

Vertical axis label: % high fibre food in the diet

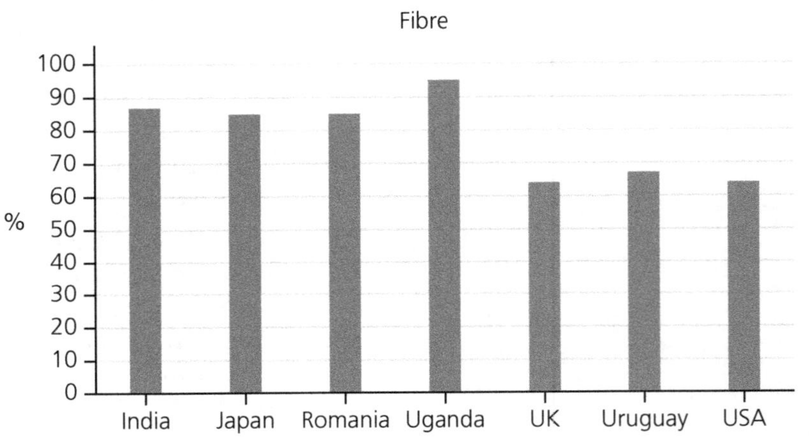

[10]

d. They do support the hypothesis – countries with very high fibre intake such as Uganda and India generally have a lower incidence of bowel cancer than countries with a low fibre intake. **[2]**

7.1 Ecosystems of the earth

Student Book pages 136–137

kerboodle

Online resource

- Worksheet 7.1.1 Different ecosystems, different habitats
- Interactive test

Prior learning

- Recognise that different animals are found in different habitats and are suited to the environment in which they live

Objective

- Identify different ecosystems on the Earth

Overview

In this lesson, you introduce your students to different ecosystems and the abiotic and biotic factors that affect them. This lays the foundations for all the ecology they will study in LCSS and in IGCSE Biology. Whenever you talk about ecosystems or habitats, discuss the **adaptations** of organisms to live there, even when you are not studying adaptations. This will help students when they consider variation, evolution and biodiversity as well as ecology.

Activities

- Read through the first paragraph on page 136 with students. Focus on your local environment as a starting point – students understand the concepts more easily when they are related to familiar settings. Then apply the ideas to other less familiar ecosystems or habitats. Ask students some of the abiotic factors that affect their own local environment – this shows you how much they remember from Chapter 4 in Stage 7. Then think about local biotic factors – the types of plants that grow locally, predators and prey species familiar to students (look for a wide range, from insects and lizards to birds or mammals), and any diseases or parasites that affect local organisms (students may not be very aware of these, so have examples ready). Students answer questions 1 and 2 on page 137.
- Read through the three land ecosystems given as examples on page 136–137. Ask students for ideas about the adaptations that living organisms will need to survive in the different ecosystems:

  Fig 7.1 A: for tropical rain forests, organisms that live underneath the trees need to survive with relatively little light.

  Fig 7.1 B: students may not realise that deserts can be hot or cold – the important factor is the lack of water, so plants and animals must have adaptations that let them survive without much water.

  Fig 7.1 C: these images show the same area of temperate woodland in summer, when it is sunny and warm and there are leaves on the trees, and in winter, when it is cold and frosty and most of the trees are bare. Organisms need the ability to survive through cold winters in temperate woodlands. Discuss different ways living things might be adapted to cope with these changes.

  Students answer Q3a on page 137.
- Read through the text on aquatic ecosystems on page 137. Students may not have thought of different aquatic environments as ecosystems before. List some of the variety in different aquatic environments. Students answer Q3b.
- Give students **Worksheet 7.1.1** and ask them to colour in the different ecosystems and make any notes they choose to about the conditions in them, ready for the next lesson.

Extension

Students investigate and write about three further ecosystems.

Homework

Workbook page

Key words

ecosystem, biosphere

7.1 Student Book answers

1. An ecosystem is an environment that supports life.
2. Abiotic factors: non living elements of an ecosystem, e.g. soil, temperature, water. Biotic: living factors in an ecosystem, e.g. predators, prey, disease.
3. **a.** Any suitable land ecosystems with main characteristics and correct level of biodiversity.
   b. Any suitable aquatic ecosystems with main characteristics and correct level of biodiversity.

7.2 Habitats within an ecosystem

Student Book pages 138–139

kerboodle

Online resources

- Worksheet 7.2.1 Animal adaptations for ecosystems and habitats
- Worksheet 7.2.2 Extension: Ecosystems and adaptations
- Interactive test

Prior learning

- Recognise that different animals are found in different habitats and are suited to the environment in which they live

Objective

- Understand that a variety of different habitats exist within an ecosystem

Overview

This lesson continues directly from Unit 7.1. Students discover that ecosystems are made up of many different habitats – and a habitat may be a very small area. As in Unit 7.1, remind students that living organisms are adapted to a particular habitat within an ecosystem where they live, and encourage them to think of as many examples as possible. Local examples help students to understand the concept. Students need to recognise that plants have habitats as well as animals.

Activities

- Ask students a few quick questions on ecosystems to make sure that they understand that concept.
- Read down to the end of the paragraph 'Where is home?' on page 138 with students. Discuss students' ideas and ask them for local examples. The key concept is that each ecosystem contains many different habitats. Organisms are adapted to the specific habitat in which they live. Students answer questions 1 and 2 on page 139.
- Work with students to consider a tree as an ecosystem containing many habitats. Use Figs 7.2 B and 7.2 C to show students some of the very different organisms that make each part of a tree their home. If possible, take students into the school grounds or a local open space to look for examples of different habitats and the organisms that live in them. Students answer Q3 on page 139.
- Students read through the content of page 139 either together or independently, thinking about different ecosystem and the habitats they contain. Remind students that dung and rotting leaves are habitats (see Chapter 3 in Stage 7).
- Students get out **Worksheet 7.1.1** from the previous lesson and are given **Worksheet 7.2.1**. Students cut out the different animal cards and match them to the correct ecosystem. Some of the animals are easier to place than others. Once students have correctly matched the adaptations to the habitats, they stick the cards in the right places. The correct matches are:

  Hot desert – A camel, B addax antelope, C fennec fox, D water-holding frog, E kangaroo rat.

  Tropical rainforest – A monkey, B flying frog, C great hornbill, D pit viper, E sloth.

  Savannah – A zebra, B impala, C giraffe, D cheetah, E termite.

  Arctic – A polar bear, B walrus, C arctic fox, D ground squirrel, E shorthorn sculpin.

  Temperate woodland – A squirrel, B stag beetle, C hedgehog, D red fox, E sparrowhawk.

  High mountain – A mountain goat, B llama, C yak, D pika, E chinchilla.

- Students answer Q4 on p139, then colour in the animals on their worksheet.

Extension

Give students **Worksheet 7.2.2** which asks them to give examples of animals living in specific ecosystems and their typical adaptations.

Homework

Workbook page 58

Key words

habitats, niche

7.2 Student Book answers

1. An ecosystem is an environment that supports life; a habitat is where an organism lives, its home.
2. Five habitats including two that are not on the spread, giving one organism that lives in each, e.g. rock pool on a sea shore – crab/shrimp/barnacle/seaweed.
3. They are all living in the same ecosystem – the tree – but they all have different habitats so they take different things from the tree, e.g. bees take nectar from the flowers, caterpillars eat the leaves, polypody ferns grow along the branches and monkeys eat fruits and leaves and move around the branches.
4. Tropical rainforest will have the most species. It is warm, wet and sunny all year round so ideal growing conditions for plants which enable huge numbers of different plant species to live there. Many animal species survive there too as they feed on the plants and each other. Alpine mountainside will have least biodiversity – very harsh conditions, very cold, water often frozen and not available – so relatively few and small plants survive which will not support many species of animals.

7.3 The interdependence of organisms

Student Book pages 140–141

kerboodle

Online resources

- Worksheet 7.3.1 Population changes on Saint Paul Island
- Worksheet 7.3.2 The story of the wolves and the caribou
- Worksheet 7.3.3 The predator–prey relationship between wolves and caribou
- Worksheet 7.3.4 Extension: Changes in the human population
- Interactive test

Objective

- Understand how organisms within an ecosystem are interdependent

Overview

Students build on their knowledge of food chains and food webs to understand how organisms are interdependent and how changes in the population of one species affects the population of other species. This lesson gives students the opportunity to use data handling skills. Students are developing a more complex understanding of the inter-relationships between organisms, which they will carry forward into their IGCSE Biology studies. This level of understanding will help them to achieve well at both CLSS and IGCSE.

Activities

- Ask students to describe a food chain and a food web. Ask them to write down three food chains and one food web.
- Discuss the interdependence of life and explain that it is more complex than simple food chains and food webs. Introduce the idea of populations – then students read page 140 of the Student Book. Give students **Worksheet 7.3.1**. This gives more detail about the case study of the deer on page 140. Students graph the population changes and suggest reasons for the changes observed.
- Discuss **Worksheet 7.3.1**. Look for explanations such as plenty of food to begin with, no predators so caribou breed fast and population grows; so many caribou that they eat all the plants → starvation, caribou can't breed, die of starvation, overcrowding means diseases spread fast, etc.
- Explain that normally there are predators and prey organisms in any ecosystem, and that in a healthy ecosystem the populations balance each other out. Give students **Worksheet 7.3.2** and **Worksheet 7.3.3**.

  Alternatively, turn **Worksheet 7.3.2** into a slideshow and go through it with students before they do **Worksheet 7.3.3**.

  OR give students **Worksheet 7.3.3** supported by the content of page 141.
- Have a plenary discussion on the interdependence of organisms and students answer questions 1 and 3. If the worksheets have not been used, they also answer Q2.

Prior learning

- Know how food chains can be used to represent feeding relationships in a habitat
- Know that food chains begin with a plant which uses energy from the sun
- Understand the terms producer, consumer, predator and prey

Extension

Students work through **Worksheet 7.3.4**. They produce a graph showing the changes in the global human population over the last 1000 years. They compare the appearance of the graph with that of the caribou on St Paul Island – it keeps increasing – and discuss factors that affect the growth of the human population, e.g. disease control through vaccination and medicines, growing our own food, importing food, controlling family size.

Homework

Workbook page 59

Key words

interdependence of life, population

7.3 Student Book answers

1. **a.** A population of animals might suddenly increase because more food becomes available/a predator competitor dies out/any other sensible point. A plant population might suddenly increase due to a big tree falling over so more light is available, a very wet season so more rain, a poor year for caterpillars – any sensible point.
   b. New disease, new predator, new competitor for food.
2. The wolf population is relatively low so the caribou population grows because there are few predators. As the caribou population grows, the wolves have more to eat and so they breed more successfully and the wolf population grows. The wolf population reaches a point when it is reducing the caribou population (there may be a lack of food for the caribou as well). The caribou population falls until there are so few caribou that the wolves die from lack of food so the caribou have few predators again and the population grows again. The two populations are interdependent and the pattern continues.
3. The interdependence of organisms reflects the way in which organisms depend on each other for survival, e.g. plants need animal waste for fertiliser to grow, herbivores need plants to eat, carnivores need prey animals. When there is little biodiversity, each species is dependent on only a small number of other species, so a change in the environment – a new disease, a drought, etc. – can have a big impact. In an ecosystem with a lot of biodiversity, there are many more species in the feeding relationships. It is much easier for organisms to simply change food source in a crisis.

7.4 Pesticides and bioaccumulation

Student Book pages 142–143

kerboodle

Online resource

- Worksheet 7.4.1 Bioaccumulation
- Interactive test

Prior learning

- Know that food chains can be used to represent feeding relationships in a habitat

Extension

Students choose a local food chain and discuss the possible consequences for the local ecosystem if a persistent pesticide was used on local crops.

Homework

Workbook page 60

Key words

pesticides, toxins, persistent pesticides, bioaccumulation

Objective

- Describe the impact of bioaccumulation of toxic substances on an ecosystem

Overview

In this lesson, you are reinforcing students' understanding of food chains and the interdependence of organisms by looking at bioaccumulation. Avoid the misconception that pesticides are a bad thing. Used mindfully, they save millions of lives by preventing the loss of food crops. Giving students a balanced view – including the problems that some pesticides cause – is valuable both for their future studies and for them as citizens.

Activities

- Ask students which animals feed on the crops we grow? Ask them to guess how much food is lost globally to pests. Describe pesticides and explain the huge benefits they give in saving food crops.
- Students read 'What are pesticides?' on page 142 and answer Q1 on page 143. Make sure that students understand the difference between a pesticide that breaks down naturally in the environment and a persistent pesticide.
- This simple activity introduces and explains bioaccumulation. Preparation: cut 25 small pieces of coloured paper and 100 small pieces of white paper. Scatter the pieces around your classroom at random. Choose 1 student to be a polar bear, and 5 students to be seals. The rest of the students are fish. Explain to your students that the coloured paper represents a persistent pesticide. The fish collect as many pieces of paper as they can find and return to their seats. Each fish totals up the number of white and coloured pieces of paper they have collected. Record each total on the board. Now the 'fish' give their paper to the nearest 'seal'. Each seal totals up the numbers of white and coloured pieces of paper they have – record these totals on the board. Finally, each seal gives its pieces of paper to the polar bear. The polar bear now has 25 pieces of persistent pesticide – enough to make it ill or even kill it. Students then complete **Worksheet 7.4.1**.
- Students read 'What is bioaccumulation?' on pages 142– 143 and answer questions 2–4.
- If there is time, students draw their own diagrams like that on **Worksheet 7.4.1** for other food chains affected by bioaccumulation.
- Summarise the lesson – make sure through questioning that students understand the value of safe pesticide use and recognise the impact of persistent pesticides both on top predators individually and on the whole ecosystem through the loss of top predators.

7.4 Student Book answers

1. **a.** A substance that kills pests.
   b. Up to 30% of the crops we grow are eaten or destroyed by pests. Farmers use pesticides to protect their crops from pests.
2. Bioaccumulation is the increase/concentration of a substance, usually a toxin, along a food chain.
3. e.g. Pesticide in water → taken in by small fish (toxin concentration low) → eaten by bigger fish (toxin concentration increases) → eaten by bigger fish (toxin concentration increases) → eaten by heron (toxin levels high enough to kill).
4. Within an ecosystem, bioaccumulation of toxic substances has many impacts. It is usually the top predator which is killed by the bioaccumulation of toxin. This means the animals that the predator eats will increase initially, and the populations lower down the chain will be reduced. In a biodiverse ecosystem the impact will be relatively low as other predators will eat the same prey, but in some ecosystems the whole feeding relationship will be destroyed.

7.5 Invasive species

Student Book pages 144–145

kerboodle

Online resource

- Worksheet 7.5.1 Invasive species poster planning
- Interactive test

Prior learning

- Investigate how different animals are found in different habitats and are suited to the environment in which they are found

Objective

- Describe how new and/or invasive species can affect other organisms and an environment

Overview

This is another lesson where you introduce new concepts to your students and at the same time reinforce existing knowledge – in this case, on the interdependence of species. Useful preparation is to have an extensive list of local examples of native, introduced and invasive species and have as many images available as possible.

Activities

- Ask students to name as many plants and animals that they have seen in their home area/country as they can. List them on the board.
- Read through page 144 (Introduction and 'What are native species?') Discuss the difference between native and non-native species and **EITHER** identify examples of both from the list you generated at the beginning of the lesson **OR** give students good local examples of each with images if possible. Remind students of the interdependence of organisms in an ecosystem and the importance of a balanced ecosystem.
- Read through the first two paragraphs of 'What are invasive species?' on page 144 with students. Make sure they understand that the problem with invasive species (compared with non-native species) is that they disrupt food chains and webs. **EITHER** identify examples of invasive species from your initial list **OR** introduce students to an invasive species that is causing problems in your country – again, images are helpful.
- Students answer questions 1–3 on page 145.
- Explain clearly to students that invasive species may be plants or animals. It can be counterintuitive to think of plants as invasive, because students think of plants as not moving far – but, in fact, some of the most damaging invasive species on Earth are plants.
- Read through the material on cane toads (page 144) and water hyacinth (page 145) with students and encourage discussion about examples of invasive species both locally and globally.
- Students complete Q4 on page 145.
- Give students **Worksheet 7.5.1**. The task is to plan a poster to warn people of an invasive species – what it looks like, why it matters and what to do about it. **EITHER** specify a local example for all students **OR** (if students have good access to the internet) allow students to choose their own example of an invasive species from anywhere in the world **OR** ask students to choose one of the examples given on pages 144–145. If possible, give students time to create their posters and then display them on the walls.

Extension

Ask students to write a piece of text to go with their poster, giving more information about invasive species in general and their featured invasive organism in particular.

Homework

Complete poster **OR** Workbook page 61.

Key words

native species, invasive species

7.5 Student Book answers

1. A native species is a natural part of the ecosystem of a country.
2. An introduced species is not a native species. It is a species that has been introduced to an ecosystem by human activity but does not change the balance of the ecosystem. An invasive species is an introduced species that damages the ecosystem it is placed in.
3. Having no natural predators in its new environment; reproducing rapidly.
4. Look for food webs similar to these. The key factors are: 1) plants at the bottom; 2) layer above cane toad in South America so it is eaten; 3) cane toad as top predator in Australia.

a.

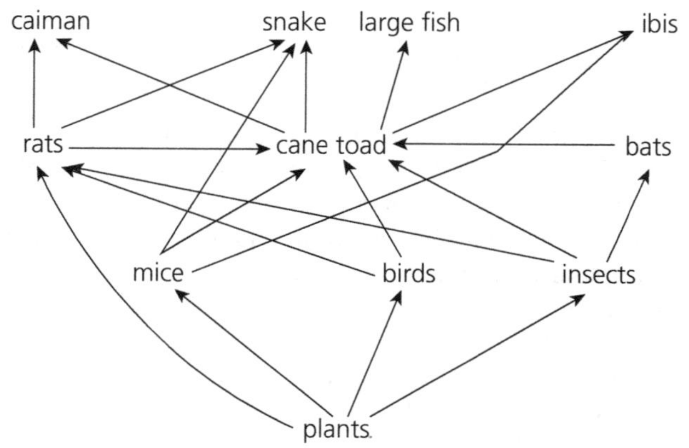

b.

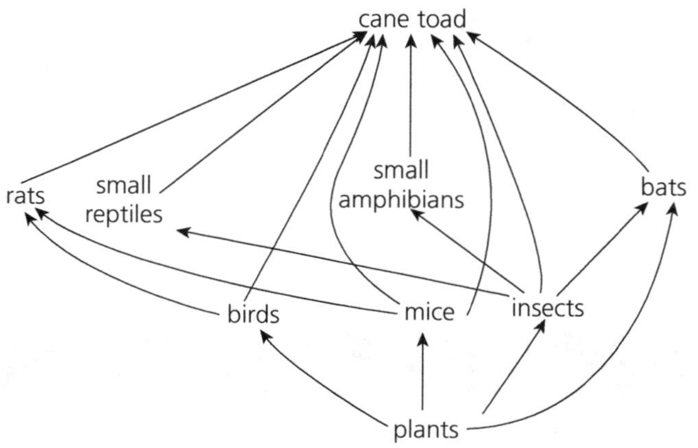

7.6 Invasive species and ecosystems

Student Book pages 146–147

Objective

- Describe how new and/or invasive species can affect other organisms and an environment

Overview

Begin this lesson with an activity. To prepare, print off sets of the cards on the Guam ecosystems for your students and cut out the sets (**Worksheet 7.6.1**). Remove the brown tree snake cards before the lesson. This lesson further reinforces many ecological concepts for students, from food webs to population dynamics and the interdependence of species, as well as the difference between native and invasive species. It provides a very practical example of the impact of invasive species on an ecosystem. You also introduce the positive side of invasive species, when they are used in biological pest control.

kerboodle

Online resources

- Worksheet 7.6.1: The ecosystems of Guam
- Interactive test

Prior learning

- Investigate how different animals are found in different habitats and are suited to the environment in which they are found

Extension

Students investigate the use of biological pest control in protecting human food crops, e.g. cottony cushion scale as a pest that damages citrus crops with vedalia beetles giving biological pest control, or ladybirds as control for aphids in greenhouses.

Homework

Workbook page 62

Activities

- Give individual students or groups of students the cards showing some of the different types of animals and plants found on the island of Guam around 60 years ago. Ask students to produce a food web for Guam using all of these organisms.
- Discuss the food webs students have developed – then tell them that around 60 years ago a non-native species was introduced to the island accidentally. Give each student/group of students a brown tree snake card. Ask them to introduce the new organism to their food web and to predict what will happen to the ecosystem in Guam.
- Students read page 146 in the Student Book to discover whether their predictions are correct. They answer questions 1–4 on page 147.
- On page 147, the Student Book covers biological pest control – you can use this as an example of all five aspects of Science in context at this level. Read through the content with students. It is important to emphasise that invasive species are rarely a problem in their own natural habitat. It is when they are introduced to a completely different ecosystem that problems arise. Prickly pear cacti are the example chosen – part of balanced ecosystems in many parts of the world but causing devastation when introduced to Australia. The solution was the introduction of another invasive species – this time a moth that fed on the prickly pear cactus but nothing else – so it did not damage local Australian species. Make sure that students are aware that biological pest control is very useful in protecting human food crops such as citrus fruits. Students answer Q5 on page 147.
- Have a plenary session summarising the work on ecosystems, habitats and the impact of different factors on populations.

7.6 Student Book answers

1. When there were fewer birds to eat, the snakes started eating lizards so they were not short of food and their population stayed high.
2. Loss of predators: Spiders were eaten by birds and lizards. The snakes have killed most of the birds and reduced the lizard population so the main predators of the spiders have gone. The spiders are not eaten so they live and reproduce and their population has increased.

   Loss of competitors for insect food: spiders competed with birds and lizards for insects as food. Without the birds and with fewer lizards there is much more food available for the spiders and so they reproduce successfully and their numbers increase.
3. They live deep in the forests and they live high up in the trees so people cannot see them easily nor can they reach them easily.
4. Insects often eat plant leaves. As the bird and lizard populations have fallen, the main predators of insects are the spiders. If the spiders are removed, the insects will have no predators. Their populations will grow and they will destroy the forests, eating the leaves and killing the trees.
5. **a.** The control of a pest species by another species of organism.

   b. It reduced the population of the pest prickly pear cactus dramatically in a short space of time. It only attacked the prickly pear – it did not damage other native species in Australia. It has maintained the balance of the ecosystem over time.

7.7 Sampling your ecosystems

TWS

Student Book pages 148–149

kerboodle

Online resources

- Worksheet 7.7.1 Investigating plant growth near a tree 1
- Worksheet 7.7.2 Investigating plant growth near a tree 2
- Interactive test

Extension

Complete questions 1–3 on page 149.

Homework

Workbook page 63

Key words

quadrat, distribution

Objectives

- Describe how new and/or invasive species can affect other organisms and an ecosystem
- Group and classify organisms through observation
- Evaluate whether measurements and observations have been repeated sufficiently to be reliable

Overview

This is an excellent opportunity to take your students outside and do some practical observations to bring back into the classroom as a basis for data handling. The Student Book gives a case study of students collecting data to identify an invasive plant, and includes a description of how to use a quadrat. The worksheets give a practical investigation, and the same investigation with data for students to work with.

Activities

- Talk to students about the need for data to show the effect of changes in an ecosystem. Read through the first three paragraphs on page 148 and demonstrate a quadrat and a random number generator to students.
- Read the section on counting samples at the bottom of page 148, along with Fig 7.7 C. You could use some paper flowers or plants – or any objects in the classroom – to demonstrate that it doesn't matter what rule you use for counting the objects in a quadrat, as long as it is always applied.
- Work carefully through the calculations on page 149 with students. Help them to understand the different sizes of quadrats and the importance of taking that into consideration when doing these calculations.
- **EITHER:** Give students **Worksheet 7.7.1**. Read through the method described, and take them outside to carry out the practical work. Ideally, pairs of students should share a quadrat. Return to the classroom to display their data as a graph. Note: you may need to add in a further calculation if your quadrats are 0.25 m^2 rather than 1 m^2 as described on the worksheet. **OR**: Give students **Worksheet 7.7.2**. Read through it describing and discussing the method used. Then students draw a graph of the results and answer the questions.
- Draw the topic to a close, reminding students that all of the ecology we learn is based on practical investigations like those they have worked on today.

7.7 Student Book answers

1. a. A measured area used to study the numbers and distribution of organisms.

b. Plants, slow-moving animals, animal evidence, e.g. droppings, nests, worm casts.

2. a. People know the approximate number of seedlings. They can pull them all up by hand. Repeating the sampling in 6 months will show if any seedlings were missed (they will be bigger) and if any new seedlings have grown. Pulling up any seedlings found again reduces the invasion of earpod wattles.

b. Repeat the sampling and pulling up seedlings at intervals for several years; try and find the adult tree that is the source of the seeds and destroy it.

3. a. To avoid bias by choosing areas that had lots of worm casts; to make their results reliable.

b. $\frac{5 + 20 + 4 + 7 + 15 + 2 + 12 + 10 + 6 + 9}{10} = \frac{90}{10} = 9$ worm casts per 0.5 m^2

c. $200 \div 0.5 = 400$ = number of quadrats that would fit on field.

Number of worm casts on the field = $9 \times 400 = 3600$ casts per 200 m^2

7.8 Review answers

Student Book pages 150–151

Student Book answers

1. a. An ecosystem is an environment that supports life. [1]
   b. Interdependent means that different types of organisms need each other to stay alive. [1]
   c. Any two from rabbit, caterpillar, snail, frog, vole, thrush. [2]
   d. Any three from frog, vole, thrush, fox, hawk. [3]
   e. It would increase – no predators to eat the rabbits so more survive and breed. [2]
   f. Caterpillar numbers would increase as they have lost a predator. Voles and thrushes would increase as there is more food for them. Foxes and hawks would hunt more voles. Foxes hunt more rabbits; hawks eat more thrushes as frogs have gone from their diet. Populations of rabbits and thrushes might fall. [5]
2. a. B [1]
   b. There is plentiful food and no predators. [2]
   c. It is falling steeply and they are dying out/the population is reducing rapidly. [2]
   d. Two from: The rabbits have eaten all the plant life on the island so there is no food and they die out/a new predator arrives on the island and eats all the rabbits/a new disease reaches the island and they all die out. [2]
3. a. The mouse population would grow as they would have no predators to eat them. [2]
   b. i. Any sensible suggestions, e.g. more plants available to the grasshoppers so their population might grow; the mouse and shrew have lost a food source so more grasshoppers might be eaten so the population would fall; the mouse has lost the slugs but gained more plant material so the mouse might not eat more grasshoppers so only the shrew would be missing food – and it has many food sources so the grasshopper might not be affected much at all. [3]
   ii. Spiders: if there are more grasshoppers, more spiders will be eaten; if fewer grasshoppers, more spiders will survive.
   Beetle population might increase because more plant food available when not eaten by slugs. If the mouse and/or shrew and/or spider populations are affected by absence of slugs, the owl population will be affected as its food sources are reduced. Any other sensible points demonstrating understanding of the interdependence of species. [5]
4. a. The increase in the concentration of a substance/toxin along a food chain. [2]
   b. Toxic spray → waterways → small fish → larger fish → bigger → heron. Annotations to include, e.g. toxic pesticide washes into river/toxins stored in fat/each animal eats many of the animal below it in the food chain/top predator absorbs so much toxin it dies. [8]
5. a. A (known) area used to study the numbers and distribution of organisms in an ecosystem. [2]
   b. Near the tree: less light because shaded by the branches and leaves of the tree: further away = more light available growing beyond the shade cast by the branches of the tree. [4]
   c. Plant B appears better adapted to dim light. Close to the tree, in the shade, it is the only plant growing. In shade with some sun by the edge of the tree branches it does extremely well. But the further out into the stronger light, plant B does less well and plant A dominates. [3]
   d. Take several readings in each place and find the mean. [1]
6. a. To avoid bias – to make her results reliable. [2]
   b. Native species – a natural part of an ecosystem; introduced species brought into an ecosystem by human activities; invasive species – an introduced organisms that damages the ecosystem it is introduced to. [3]
   c. i. Year 1: In a $0.5\,m^2$ quadrat = $(1 + 2 + 0)/3 = 3/3 = 1$ plant per $0.5\,m^2$; $50/0.5 = 100$; $1 \times 100 = 100$ strange plants per $500\,m^2$. [3]
   ii. Year 2: In a $0.5\,m^2$ quadrat = $(6 + 4 + 5)/3 = 15/3 = 5$ plants per $0.5\,m^2$; $50/0.5 = 100$; $5 \times 100 = 500$ strange plants per $500\,m^2$. [2]
   iii. Year 3: In a $0.5\,m^2$ quadrat = $(30 + 25 + 20)/3 = 75/3 = 25$ plants per $0.5\,m^2$; $50/0.5 = 100$; $25 \times 100 = 2500$ strange plants per $500\,m^2$. [2]
   d. Selina's concerns appear to be correct. The population of the strange plant is growing fast. It does not appear to have predators. However she is only doing 3 quadrats at a time and only once a year so the data is too limited to draw a reliable conclusion. [3]
   e. Do more quadrats; do them more often during the year; try to identify the plant to discover if it is an invasive species; any other sensible point. [2]

8.1 What is weather

Student Book pages 152–153

kerboodle

Online resources

- Worksheet 8.1.1 Weather symbols
- Worksheet 8.1.2 My weather report
- Interactive test

Extension

Give students the weather forecasts for the previous three days that you have recorded and ask them to assess how accurate they were and give possible reasons why they were so accurate/so inaccurate.

Homework

Workbook page 64

Key words

weather, climate

Objective

- Describe weather

Overview

In this chapter, you will introduce your students to the differences between weather and climate, and explain what is meant by climate change. In this first lesson, your students learn about weather as a mix of atmospheric events taking place at a particular place at a particular time. You should emphasise that weather is extremely local. For use in an extension exercise, note down or record the weather forecasts for three days before your lesson.

Activities

- Ask students for examples of different kinds of weather they have experienced or heard about. Write a list on the board – aim to get as many different types of weather as possible.
- Read through page 152 'What is weather?'. Link weather to the abiotic factors that affect our environment. Help students to understand that the weather we experience on the ground is the result of changes in the atmosphere above us. Students answer questions 1 and 2 on page 153.
- Read through the section titled 'Predicting the weather' with students. Talk about local and national weather forecasts – if possible, show students examples of each. Discuss the importance to different industries of having an idea of the weather that is expected.
- Give students **Worksheet 8.1.1** and ask them to complete the table, identifying the weather shown by each symbol. Some will be familiar to them – others may be less familiar.
- Students read through page 153 and answer questions 3 and 4.
- Give students **Worksheet 8.1.2**. Depending in the time left in the lesson, **EITHER** ask students to think back over the last week and fill in as much detail about the weather each day as they can remember, ending on the day of your lesson **OR** ask students to start on the day of your lesson and fill in a detailed weather report for each day of the following week. Illustrations with notes and/or weather symbols are good ways to do this.

8.1 Student Book answers

1. Weather is a mix of atmospheric events that happen in a particular place at a particular time.
2. Two from: Atmospheric pressure; humidity; air temperature.
3. Correctly drawn and identified weather symbols from text book or local weather app or similar.
4. **a.** Draws on recordings of atmospheric pressure, rainfall, hours of sun, speed and direction of wind from weather stations all over the country over time to predict the weather to come.

   b. Weather can be very local, affected by local features such as hills or lakes, and atmospheric conditions can change very fast and so national forecasts are sometimes wrong.

8.2 Climate and climate change

Student Book pages 154–155

kerboodle

Online resource

- Worksheet 8.2.1 Global climates
- Interactive test

Extension

Students do a thought experiment and imagine the impact on their own country if the climate became a) hotter and b) colder.

Homework

Workbook page 65

Objectives

- Describe the difference between climate and weather
- Understand that the climate of the Earth changes over time

Overview

This unit continues directly from Unit 1.1 in which students learned about the weather. Now you introduce the concept of climate – and compare the two. Be prepared to use experiences students may have had of different climates, or tell anecdotes from your own travels to different climates, if relevant. It can be hard for students to imagine how different it is. If you give your students a clear understanding of the distinction between weather and climate, it will help them to achieve well in their future biology studies and avoid common misconceptions.

Activities

- Look out of your classroom window and comment on the conditions – sunny, windy, rainy, etc. Now ask your class – are you describing the climate or the weather? Remind them that weather describes the local conditions now – so they are describing the weather.
- Read page 154 of the Student Book with students, reinforcing the concept that climate is measured over many years and describes long-term weather patterns in an area, and look at the different global climates described in the table on pages 154–155. Decide with students what type of climate you all live in. Ask students for experiences of visiting or living in different climates.
- Give students **Worksheet 8.2.1** and ask them to fill it in, based on the table in their Student Books. Ask them to name and describe the climate in the left-hand column and draw something that represents the climate in the right-hand column. They should NOT simply draw the photographs – they are to think of plants, animals, food or some other aspect of the climate that sums it up, e.g. a tropical climate could be a palm tree, a pineapple, or a lizard.
- Give students the statement **Climate is what you expect; weather is what you get** and then ask them to read through the section 'What is the difference between weather and climate?' on page 155.
- Introduce the concept of climate change as happening but as a slow process. Read the final section on page 155 with students and discuss Fig 8.2 A as evidence – it shows a trilobite, an animal that lived in tropical seas millions of years ago. Fossilised tropical shells are found in similar places – where the climate is now cool and temperate. This shows that the climate changes over thousands or millions of years.
- Students complete questions 1–3 on page 155.
- Finish with a brief plenary to confirm that students have a clear understanding of weather and climate and the differences between them.

8.2 Student Book answers

1. Climate describes the long-term weather patterns in an area.
2. Weather is a mix of atmospheric events that happen in a particular place at a particular time. Climate describes the weather you would expect in a particular area based on recordings over a long period of time.
3. Three correctly identified climatic zones and the main weather patterns associated with each.

8.3 Climate changes past and present

Student Book pages 156–157

kerboodle

Online resources

- Worksheet 8.3.1 Adaptations to climate
- Worksheet 8.3.2 Carbon dioxide concentration and temperature at the surface of the Earth over time
- Interactive test

Objectives

- Understand that there is evidence that there are cycles of climate change over long periods of time
- Understand that climate change results from atmospheric change

Overview

A common misconception with students is that climate change is a new thing. In this lesson, you reinforce understanding that the climate of the Earth has changed many times through history. The Student Book provides strong images and data – which is also used in the accompanying worksheet. Helping students understand the factors that affect natural climate change at this stage of their learning journey prepares them for future content on climate change in Chapter 12 and supports them through to IGCSE Biology and beyond, where they have to demonstrate clear understanding of both natural and anthropogenic climate change.

Activities

- Brainstorm students to discover what they know – or do not know – about climate change. Build a web of statements on the board.
- Read through page 156 and the first paragraph of page 157 with students, looking at the different figures to illustrate the content. Many students will have heard of dinosaurs or mammoths – they may not have realised that they lived in very different climates. Give students **Worksheet 8.3.1** and give them a few minutes to list some ideas. Suggest that they think about adaptations for survival in different conditions – and remind students that dinosaurs were reptiles, with limited ability to control their own body temperature, but mammoths were mammals. Then go through some of the ideas, asking for points from different students and giving students time to complete their lists when new points are added.
- Students answer Q1 on page 157.
- Read through the content on 'Natural cycles of climate change' with students and give them **Worksheet 8.3.2**, which provides a bigger version of Fig 8.3 E for them to use when answering questions 2 and 3 on page 157.
- Read through the final section 'Climate change today' with students. Complete a quick calculation to show how much faster climate change is happening now than it did historically. Have a brief discussion about the impact of human activities and mention that there is more learning to come on this in Stage 9.
- Have a plenary discussion on climate change – for example, see how many of their initial ideas were correct – and highlight what has been learned. Challenge students to find out about where the evidence comes from before the next lesson.

Extension

Students research conditions during the age of dinosaurs and the time of the mammoths – ask them to find out how long ago each lived, the average temperature at the surface of the Earth at the time, the level of carbon dioxide in the atmosphere at the time, etc.

Homework

Workbook page 66

Key word

fossils

8.3 Student Book answers

1. Every 100 000 years, the Earth moves further from the Sun in its orbit, cooling the climate.

   Changes in the activity of the Sun mean it is hotter at some times than others, warming or cooling the Earth.

   Volcanos erupting/big meteorites hitting the Earth fill the atmosphere with ash and dust. This stops sunlight reaching the Earth and the climate cools down.

   Carbon dioxide levels in the atmosphere go up and down. Volcanoes and living things release carbon dioxide. Plants take it in. Carbon dioxide traps energy from the Sun, keeping the surface of the Earth warm enough for life. High carbon dioxide levels mean the climate warms, and low carbon dioxide levels cause cooling.
2. **a.** A regular pattern of the climate of the Earth cooling and then warming over hundreds of thousands of years.

   b. There are periods when the Earth cools down. They are called ice ages and they last 80–90 000 years. The temperature is about 8 degrees below the average. Then the Earth warms up and goes into a warm period for 10–20 000 years when it is about 2 degrees above the average before it cools down and enters another ice age.

   c. The concentration of carbon dioxide in the atmosphere shows a similar pattern of rises and falls as the temperature. They appear very closely linked. It seems probable that rises in carbon dioxide are linked to global warming and falls due to global cooling.
3. **a.** Rate of warming since last ice age = 5 °C per 5000 years so rate = 5000/5 = 1 °C per 1000 years.

   Rate of warming over last 100 years = 1 °C per 100 years.

   b. Rate of warming from last ice age follows the natural cycle. The rate of warming over the last 100 years is 10 times faster that the natural rate of warming, supporting the idea that the climate is now changing faster than the natural cycle.

8.4 Gathering evidence of climate change

SIC

Student Book pages 158–159

kerboodle

Online resources

- Worksheet 8.4.1 Carbon dioxide in the atmosphere
- Worksheet 8.4.2 Extension: Data from Antarctic ice cores
- Interactive test

Objectives

- Understand there is evidence for the natural cycles of climate change on Earth
- Understand that climate change results from atmospheric change
- Discuss how scientific knowledge is developed through scientists working together over time
- Describe how people develop scientific understanding

Overview

This lesson helps students to develop their understanding of Science in context. It is easy to present students with facts, but it is very important that they understand where the information comes from, how data is collected and how scientists work together internationally to develop ideas. This lesson gives you information to share with students on ways of monitoring climate change now, and techniques used to measure changes in the atmosphere and the climate over thousands of years.

Activities

- Have a quick question and answer session to confirm that students remember and understand the work so far on weather, climate and climate change.
- Ask students how they think scientists measure changes in the climate that are happening now. Then read through to the end of 'Observing changes now' on page 158. Students answer questions 1 and 2 on page 159.
- Ask students to think about how scientists might be able to measure what the carbon dioxide concentration in the atmosphere or the temperature at

the surface of the Earth were, many thousands of years ago. They should suggest using fossil evidence as they have met this before. Read through 'Understanding past changes' on pages 158–159, using Figs 8.4 C–E to describe the value of fossil evidence, ice cores and pollen records, before students answer Q3 to reinforce their learning.

- Give students **Worksheet 8.4.1**. This is a bigger version of Fig 8.4 A. Students are asked to read data from the graph and make a bar chart to show the trend in atmospheric carbon dioxide levels. This reinforces the findings and also gives them the opportunity to carry out data manipulation.
- Summarise the lesson and its relevance to the topic as a whole in a plenary session before students revise the content for their review test.

Extension

Give students **Worksheet 8.4.2**. They represent data from Antarctic ice cores as a bar chart and discuss the results in terms of climate change and the concentration of carbon dioxide in the atmosphere.

Homework

Workbook page 67

8.4 Student Book answers

1. a. Peer review is the process by which scientists check each other's work before it is published.

b. Peer review makes sure the research is reliable before it is published.

2. e.g.

Method	Main findings
Analysing carbon dioxide content of the air	Increasing every year
Measuring the temperature of the surface of the Earth in different places	Temperatures are rising steadily
Recording extreme weather events	They are increasing in number and becoming more extreme
Recording ice cover in polar regions	Ice melting more each year

3. a. It is hard to measure the temperature or the changes in gases in the atmosphere thousands of years ago.

b. i. They look at fossil animals and plants/pollen grains which are found deep in ice or bogs. The type of animals or plants identified tell scientists about the temperature conditions when they were alive.

ii. Scientists drill down and remove ice cores – the ice is thousands of years old. Ice contains air. The air in the ancient ice cores is also very old. Scientists analyse the air trapped in the ice to find out what the concentration of the different gases was.

8.5 Review answers

Student Book pages 150–151

Student Book answers

1. a. Weather is a mix of atmospheric events taking place at a particular time and a particular place. **[2]**
   b. Any three sensible suggestions, e.g. sunny, rainy, windy. **[3]**
   c. Two from humidity/atmospheric pressure/air temperature. **[2]**
2. a. i. thunderstorm = D
   ii. sunshine = A
   iii. rain = B
   iv. snow = E
   v. cloud = C **[5]**
   b. A prediction of what the weather will be like based on observations by scientists of factors such as atmospheric pressure, temperature, wind, etc. **[3]**
   c. Farmers – to know when to plant seeds/harvest crops; food manufacturers – to know what type of food to make in big quantities. Any other sensible suggestion with reason. **[4]**
   d. Weather is a very local thing – it can be different over a very small area. Features like hills, mountains and lakes affect the weather, so it is very hard to get the forecast correct for every place. **[3]**
3. a. Both involve factors such as wind and rain, etc. Weather is a mix of atmospheric events taking place at a particular time and a particular place. Climate describes the long-term weather patterns in an area. **[4]**
   b.

Climate zone	Conditions
Arid	Very dry, very little rain
Continental	Short, hot summers, long, cold winters
Mediterranean	*Hot, dry summers, cool wet winters*
Polar	Extremely long, very cold winters, short cold summers
Temperate	Mild summers and cool winters
Tropical	Hot and humid, lots of rainfall e.g. Caribbean countries

[10]

4. a. Natural cycles of climate change. **[1]**
   b. The climate cools down for a period of many thousands of years – there is an ice age. **[2]**
   c. The temperature falls to around 8 degrees lower than the average – much of the land and sea covered by ice – plants and animals struggled to survive. **[2]**
5. a. The position of the Earth to the Sun in its orbit changes/over thousands of years. When the Earth moves further away from the Sun in its orbit, the Earth cools/when it is nearer to the Sun, the Earth warms. **[3]**
   b. The activity of the Sun changes. When it is most active, the Earth gets warmer/when it is least active, the Earth cools. **[2]**
   c. Volcanic eruptions fill the atmosphere with dust and ash. Blocks sunlight reaching the surface – the Earth cools. **[3]**
   d. Carbon dioxide levels in the atmosphere go up and down. Volcanoes and living things release carbon dioxide. Plants take it in. Carbon dioxide traps energy from the Sun, keeping the surface of the Earth warm enough for life. High carbon dioxide levels mean the climate warms, and low carbon dioxide levels cause cooling. **[5]**

6. a. Takes 5000 years to increase by 5 °C.
To increase by 1 °C 5000/5 = 1000 years. **[3]**

b. Current heating 1 °C every 100 years.
To increase by 5 °C = 100 × 5 = 500 years. **[3]**

c. Natural cycle 1 °C in 1000 years. Modern warming 1 °C in 100 years. 1000/100 = 10. So the rate of climate warming is 10 times the normal, natural rate.

OR Natural cycle 5 °C in 5000 years. Modern warming 5 °C in 500 years. 5000/500 = 10. So the rate of climate warming is 10 times the normal, natural rate. **[4]**

7. a. 319 ppm (±1) **[1]**

b. 370 ppm (±1) **[1]**

c. 370 ppm in 2000, 390 ppm in 2010, so an increase of 20 ppm between 2000 and 2010. **[1]**

d. The temperature will increase. Carbon dioxide traps energy from the Sun and warms the surface of the Earth. More carbon dioxide in the atmosphere will mean more warming. **[5]**

8. a. Fossils are the remains of animals and plants from millions of years ago. Scientists identify the organisms. They know the conditions they need to live. This tells them the climate at the time of the fossils. Pollen lasts thousands of years in bogs and ice. The pollen from each type of plant is different. Scientists can identify plants from ancient pollen. They type of plant tells them about the climate at the time the plant was alive. **[6]**

b. i. When the work of scientists is checked by other scientists (before publication). **[1]**

ii. It makes sure that a piece of research is reliable and can be repeated before it is published as a finding. **[2]**

iii. It allows scientists to share ideas/and how to do experiments (experimental methods)/with other scientists all over the world. **[2]**

Stage 8
Review answers

Student Book answers

1. a. glucose + oxygen → carbon dioxide + water + energy **[3]**
   b. The mitochondria **[1]**
   c. Cells need energy to carry out all the processes of life. They cannot use the energy in food directly. Aerobic respiration releases the energy cells need in a controlled way so all cells must to carry out the process. **[2]**
2. a. When the body exchanges oxygen from the air for carbon dioxide in the blood in the alveoli of the lungs. **[2]**
   b. The reactions that release energy from food in a controlled way in the cells. **[2]**
   c. Moving air in and out of the lungs by movements of the ribs and diaphragm. **[2]**
3. a. The site of gas exchange in the lungs **[1]**
   b. Any three from:
   - Good blood supply – surrounded by a network of tiny blood vessels called capillaries. Brings plenty of oxygen-rich blood to the alveoli so gas exchange can take place.
   - Huge numbers of alveoli give a big surface area for gas exchange.
   - Steep concentration gradient for diffusion of gases between the air in the alveoli and the blood in the capillaries. Maintained by breathing movements bringing air in and out and heart pumping blood through the blood vessels.
   - Thin walls so short diffusion distances between the air in the alveoli and the blood. **[6]**
4.

Blood component	Function
Red blood cell	**Carry oxygen from lungs around the body**
White blood cells	**Fight disease**
Plasma	**Carries blood cells, dissolved food, waste substances, hormones, etc.**
Platelets	Help clots to form if you cut yourself

[4]

5. a. A balanced diet means taking in a variety of foods that supply all the nutrients you need to keep you healthy. **[2]**
   b. Carbohydrates, proteins and lipids/fats and oils. **[3]**
   c. Minerals. **[1]**
   d. Spinach has 25 mg of vitamin C per 100 g. We need 75 mg of vitamin C. 75/25 = 3 so 100 × 3 = 300 g spinach per day. **[3]**
   e. i. e.g. Vitamin A in spinach = 0.45 mg/100 g, so 200 g spinach will give 0.45 × 2 = 0.9 mg vitamin A or 100 g carrot gives 0.80 mg vitamin A and 100 g spinach has 0.45 mg, so 100 g each of carrot and spinach would give more than the daily requirement. Any one of a number of ways, but must be equal or more than the actual amount. However, 100 g banana gives only 0.003 mg, so would need 3 kg banana (300 × 100 g) to give 0.9 g vitamin A. Similarly, would need 75 × 100 g orange (7.5 kg), so not realistic. **[4]**
   ii. She would have to eat very large amounts of them to get the vitamin A she needs. **[1]**
   iii. Alternative food sources – one from: egg yolk, fish liver oils, dairy products, green vegetables. **[1]**

6. a. The mass of air in a single breath. [1]
   b. The size of the balloon/the fact they each do a single breath/age of the participants. [1]
   c. They show that the mean mass of air breathed out in one breath by smokers was 31 units and by non-smokers of the same age was 40 units. [2]
   d. Tar lines the alveoli and makes them less efficient at gas exchange. Tar and other chemicals in cigarette smoke are carcinogenic and increase the risk of developing cancers of the respiratory tract/lungs, etc. Cigarette smoke paralyses the cilia in the respiratory tract so they do not clear away bacteria, which makes infections of the respiratory tract more likely. [7]
7. a. i. An environment that supports life. [1]
   ii. Home for specific organism. [1]
   iii. The variety of different types of organisms in an ecosystem. [1]
   b. Any **two** valid land ecosystems and **one** aquatic ecosystem. [3]
   c. Must be two habitats found in the chosen ecosystem. [2]
   d. Producer – plants, organisms that make food by photosynthesis. Predator – an organism that preys on and eats other animals. Parasite – feeds on another living organism and weakens it. [3]
   e. An invasive species is non-native; it has been introduced to a country or ecosystem by people. [2]
   f. Invasive species have no predators or animals that feed on them. They often reproduce quickly. As a result, their population grows rapidly and they quickly affect other organisms, breaking existing food webs and feeding relationships, etc. [4]
8. a. To avoid bias if students chose where to put them. [1]
   b. Divide the areas into grids and use a random number generator to determine the coordinates for their quadrat. [2]
   c. Mean worm casts in trampled area = 4 + 3 + 7 + 2 + 2 = 18: 18/5 = 3.6
   Mean worm casts in flower bed = 6 + 7 + 9 + 9 + 8 = 39; 39/5 = 7.8 [4]
   d. Worms do not survive well in/do not like trampled earth so they die or move away/ worms grow and reproduce in well-composted soil so their numbers increase. [2]
9. a. Weather is local. It is the mix of atmospheric events taking place at a particular place at a particular time. Climate describes the long term weather patterns in an area. [3]
   b. Any three from:
   - Changes in the carbon dioxide concentration in the atmosphere.
   - The position of the Earth on its orbit around the sun.
   - Changes in the activity of the Sun.
   - Events such as volcanic eruptions/meteorites hitting Earth. [3]

   c. Measuring carbon dioxide levels in the atmosphere/measuring temperature at surface or the Earth/in the seas/measuring the extent of ice cover or glaciers/any other sensible measure. [2]
   d. Ice cores, pollen, fossil evidence. [3]

9.1 What do we know about plants?

Student Book pages 164–165

kerboodle

Online resources

- Worksheet 9.1.1 The structure of a plant
- Worksheet 9.1.2 How people use plants
- Interactive test

Prior learning

- Know that plants have roots, stems, leaves and flowers

Extension

Students complete Q3c and d on page 165.

Homework

Workbook page 68

Key word

yield

Objectives

- Describe the structure of plants from the roots to the leaves
- Know that photosynthesis is the process by which plants make carbohydrates
- Describe trends and patterns in results
- Make conclusions by interpreting results

Overview

This is the first lesson of eight in which students learn more about plants and the processes they carry out. Discover what your students remember from work on plants in Chapter 2 (in Stage 7), for example, and on food chains and webs. One major focus of this lesson is to reinforce the importance of plants to people. The other is to introduce the importance of plant scientists and the impact they make on global health and well-being.

Activities

- Give students **Worksheet 9.1.1**. Ask them to colour in the different parts of the plant and then to add their names and notes on their functions in pencil.
- Read through 'The structure of a plant' on page 164 with students and ask them to revisit their diagrams and see how accurately they have labelled the parts. Students **EITHER** label the plants correctly in ink (rubbing out their initial pencil notes) **OR** answer Q2 (page 165).
- Challenge students to list as many ways as possible in which plants are important to humans. Then read through the section headed 'Why are plants important?' with students – check whether they thought of all the uses listed and discover if any of them thought of further uses. Give students **Worksheet 9.1.2**. They complete the table, giving as many specific examples of plant products and how they are used by people as possible, e.g. wood: building material, fuel, carving or jewellery, furniture, utensils.
- Now introduce the importance of plant scientists and the work they do, working through the section 'Why do we study plants?' with students before looking at the Thinking and working scientifically box as an example of what can be done.
- The Thinking and working scientifically box gives students information on the way scientists have increased the yield of different crops. It includes data on the increased yields in a range of crop plants, many of which students will be familiar with. Work through the content with students, emphasising ways to describe and interpret the results shown in Fig 9.1 B and some of the conclusions that can be drawn before students work through Q3a and b on page 165.
- Draw together all the main points from the lesson.

9.1 Student Book answers

1. As food; as part of the water cycle: as medicines; any other sensible points.
2. Drawing like Fig 9.1.C with added flower and label to flower. Roots – anchor the plant and supply water and nutrients. Stem – supports plant. Leaves – capture energy from the Sun in photosynthesis. Flowers – reproduction.
3. **a.** The amount of every crop produced has increased over time.
   b. **i.** Potatoes
   ii. 1970: 14.35 tonnes per hectare; 2017 21.21 tonnes per hectare
   iii. $21 - 14 = 7 \rightarrow 7/14 \times 100 = 50\%$
   c. Cassava : $11.5 - 8.5 = 3 \rightarrow 3/8.5 \times 100 = 35.3\%$
   Rice: $4.5 - 2.5 = 2 \rightarrow 2/4.5 \times 100 = 44.4\%$
   Maize: $6 - 2.5 = 3.5 \rightarrow 3.5/6 \times 100 = 58.3\%$
   d. Scientists have investigated plant structures, plant breeding, plant genetic material, the nutrients plants need and the pests and diseases that attack them. It is important because it has allowed them to develop plants with useful characteristics that produce much more food than ever before.

9.2 Photosynthesis

Student Book pages 166–167

kerboodle

Online resources
- Worksheet 9.2.1 The word summary equation for photosynthesis
- Worksheet 9.2.2 The structure and function of a plant leaf
- Worksheet 9.2.3 Extension: Comparing aerobic respiration and photosynthesis
- Interactive test

Prior learning
- Know that plants need energy from light for growth
- Explain observations that plants need water and light to grow

Extension

Give students **Worksheet 9.2.3**. This challenges them to remember the word equation for aerobic respiration and compare the processes of respiration and photosynthesis – moving them towards the level of understanding required for IGCSE Biology.

Homework

Complete Q4 on page 167 and/or Workbook page 69.

Key word

biomass

Objectives

- Know that photosynthesis is the process by which plants make carbohydrates, using the energy from light
- Know that photosynthesis occurs in the chloroplasts
- Know and use the summary word equation for photosynthesis

Overview

This lesson continues directly from Unit 9.1. Students become familiar with the basic biochemistry of photosynthesis, the importance of chloroplasts in the process and the way the products of photosynthesis are used. Students need a clear understanding of photosynthesis and its importance to plants to achieve the highest grades in IGCSE Biology, so a strong foundation built in this and the following lessons will help them do well both now in CLSS and in the future.

Activities

- Use a quick quiz to make sure that students remember the different ways in which we use plants from the previous lesson.
- Remind students that – like aerobic respiration – photosynthesis is really lots of chemical reactions that we simplify to one process at this stage of biology. Read page 166 with students, pausing to discuss and explain the different aspects of photosynthesis given in the text. Look at Fig 9.2 C and ask students to think about the different substances moving into and out of a leaf.
- Give students **Worksheet 9.2.1**. They write out the word summary equation and decorate it to help fix it in their memory. Then students answer Q1 on page 167.
- Read through the paragraph about chloroplasts at the top of page 167. Make sure your students are clear that not all plant cells contain chloroplasts. Any part of the plant that is not green, e.g. root cells, do not contain chloroplasts. The epidermis cells that form a protective outer layer to the leaf do not contain chloroplasts – they are transparent to let the light through. Students answer Q2 on page 167.
- Students read through the final section on page 167 about the ways in which plants use the products of photosynthesis and answer Q3.
- Give students **Worksheet 9.2.2**. Students must use what they know about the structure of the leaf and different types of plant cells, what they have learned about photosynthesis and Fig 9.2 C to help them to label a diagram of a section of a plant leaf. If students complete **Worksheet 9.2.2**, they answer Q4 on page 167.

9.2 Student Book answers

1. a. Photosynthesis is the process by which plants make their own biomass/food using carbon dioxide and water and light energy captured by chlorophyll.

   b. $$\underset{\text{(reactants)}}{\text{carbon dioxide + water}} \xrightarrow[\text{chlorophyll}]{\text{light}} \underset{\text{(products)}}{\text{glucose + oxygen}}$$

2. Chloroplasts contain chlorophyll; the green colour/pigment that traps light energy for photosynthesis. Chloroplasts are also the site of most of the reactions of photosynthesis. Without chloroplasts there is no photosynthesis.
3. For aerobic respiration/as a starch store/to make other molecules such as proteins, etc.
4. Plants make their own biomass by photosynthesis and a lot of the biomass on the Earth is plants. Most other organisms depend on eating plants, or on eating animals that have eaten plants, to get their biomass – so directly or indirectly all of their biomass also comes from plants. The mass of decomposers also comes from breaking down plants or animals. So most of the biomass on Earth comes from photosynthesis.

9.3 Evidence for photosynthesis: testing for starch

TWS

Student Book pages 168–169

kerboodle

Online resource

- Worksheet 9.3.1 Testing a leaf for starch
- Interactive test

Prior learning

- Plants need energy from light for growth

Objectives

- Know that photosynthesis occurs in the chloroplasts, and is the process by which plants make carbohydrates, using the energy from light
- Plan a range of investigations of different types to obtain appropriate evidence when testing hypotheses

Overview

This lesson helps students develop to their skills in Thinking and working scientifically.

This is a good opportunity for students to carry out practical work themselves, as well as evaluating different practical methods described. Relatively little apparatus is needed if you provide your students with boiling water from a kettle rather than them boiling water using a Bunsen burner. An alternative is to demonstrate the basic technique of testing a leaf for starch. Forward planning is required to have one plant in a dark cupboard for at least 3 days before the lesson, and another plant in light for at least 12 hours before the lesson.

This is a valuable opportunity to demonstrate to students how different types of investigations give different evidence about the same process. It may also demonstrate the variability of living organisms.

Activities

- Begin the lesson with quick questions on what a plant needs to photosynthesise. Introduce the rest of the lesson as a practical session/ evaluation of different practical techniques to investigate or demonstrate what is needed for photosynthesis.
- **EITHER** give students **Worksheet 9.3.1** for starch and let them carry out this practical technique themselves **OR** demonstrate the practical technique given on **Worksheet 9.3.1** to the class.
- Bring the class together and discuss how the basic technique – testing a leaf for the presence of starch as a sign that photosynthesis has taken place – can be used to demonstrate what a plant requires to photosynthesise successfully. Students answer Q1 on page 169, producing a flow diagram of the practical technique.
- Students work through 'Investigating photosynthesis' on page 169 and answer Q2 in full.
- Have a plenary session to go through the answers to Q2 and bring together ideas on the strengths and weaknesses of these techniques for demonstrating or investigating photosynthesis.

Extension

Students write a paragraph to explain why it is much harder to demonstrate that plants need carbon dioxide and water for photosynthesis than that they need light and chlorophyll. [You can remove carbon dioxide from the air around the leaf but can't stop the leaves making carbon dioxide as the cells respire, so they always have some carbon dioxide; plant cells contain lots of water and if a plant has no water input it will wilt and die, so you can't investigate the effect of removing water].

Homework

Workbook page 70

9.3 Student Book answers

1. Take a leaf from a test plant → place it in boiling water for up to a minute (to remove waterproof covering/break open cells) → turn off heat → remove leaf from boiling water → place leaf in test tube of ethanol → place test tube of ethanol in beaker of hot water (ethanol boils/removes green colour from leaf) → remove pale, stiff leaf from ethanol → dip leaf into hot water (to soften) → spread leaf on white tile → add a few drops of iodine solution (to test for starch) → if starch is present, turns blue-black.
2. **a.** Plants turn some of the glucose they make during photosynthesis into starch in their leaves to use when they are in the dark. It takes two to three days to use up these starch stores. By using plants that have been kept in the dark, students know that any starch they find in their tests is the result of photosynthesis during their investigation.

   b. Salma: **i.** Whether chlorophyll is needed for photosynthesis.
   - **ii.** Chlorophyll is needed for photosynthesis.
   - **iii.** Any two from these or any other sensible point: using a plant that has been kept in the dark, using a variegated leaf, leaving the plant in the light for 24 hours, using the iodine test for starch.
   - **iv.** Use more than one leaf of each type; use green and white leaves instead of green and pink leaves; try several different types of variegated leaves; any other sensible point.

   Dina: **i.** Whether light is needed for photosynthesis.
   - **ii.** Light is needed for photosynthesis.
   - **iii.** Any two from these or any other sensible point: using a plant that has been kept in the dark, covering part of a leaf with foil/card so the light couldn't reach it, leaving the plant in the light for 24 hours, using the iodine test for starch.
   - **iv.** Partly cover more than one leaf; completely cover some leaves; use more than one plant; any other sensible point.

   Abdul: **i.** Whether carbon dioxide is needed for photosynthesis.
   - **ii.** CO_2 is needed for photosynthesis.
   - **iii.** Any two from these or any other sensible point: using a plant that has been kept in the dark, removing the carbon dioxide from around one leaf, using the iodine test for starch.
   - **iv.** Use more than one leaf of each type; use green and white leaves instead of green and pink leaves; try several different types of variegated leaves; any other sensible point.

9.4 Evidence for photosynthesis: oxygen bubbles

TWS

Student Book pages 170–171

Objectives

- Know that photosynthesis occurs in the chloroplasts, and is the process by which plants make carbohydrates, using the energy from light
- Decide when to increase the range of measurements taken to give reliable data
- Describe trends and patterns in results
- Make conclusions by interpreting results and explain the limits of your conclusions

Overview

This lesson helps students to develop their skills in Thinking and working scientifically. It is good opportunity for practical work with your students or a practical demonstration that requires very little apparatus. The Student Book provides a version of the practical and results for students to manipulate. Students already know that plants need light for photosynthesis. In this lesson, they investigate the link between light intensity and the rate of photosynthesis. This follows up on earlier work on the need for different techniques to investigate different aspects of the same process. It also introduces the limitations of the iodine test for starch as a way of assessing the rate of photosynthesis.

kerboodle

Online resources

- Worksheet 9.4.1 How does light intensity affect the rate of photosynthesis?
- Worksheet 9.4.2 The effect of light intensity on the rate of photosynthesis: Thinking about variables
- Worksheet 9.4.3 Extension: How does light intensity affect the rate of photosynthesis? Looking at the data
- Interactive test

Prior learning

- Plants need energy from light for growth

Extension

Students complete **Worksheet 9.4.3**.

Homework

Workbook page 71

Key word

light intensity

Activities

- Ask students whether they think a plant would photosynthesise faster in bright light or dim light. Ask what the term 'rate of photosynthesis' means. Encourage students to suggest ways in which the rate of photosynthesis might be measured.
- Discuss problems using the iodine test for starch in measuring the rate of photosynthesis and point out its limitations – especially the fact that the plant leaves have to be killed before you can test them. Students then read the section headed 'Evidence for photosynthesis' on page 170.
- Read through the section headed 'Does light intensity affect the rate of photosynthesis?' to the end of the method on page 171.
- **EITHER** give students **Worksheet 9.4.1** and ask them to complete the sections headed Apparatus, Method and Prediction. As part of this process, they also complete **Worksheet 9.4.2**.

  Students carry out the practical as described on pages 170–171 and complete the results table on **Worksheet 9.4.1**. They use this data to complete the worksheet, displaying the results on a line graph, drawing conclusions and evaluating what they have done.
- **OR** give students **Worksheet 9.4.1** and **Worksheet 9.4.2**. Discuss the practical with students and get them to complete **Worksheet 9.4.2**, considering the variables involved. Carry out the experiment as a demonstration, with students completing the **Worksheet 9.4.1** sections headed Apparatus, Method and Prediction whilst the apparatus equilibrates each time. Students use the data from your investigation or the data on page 171 to complete the worksheet, displaying the results on a line graph, drawing conclusions and evaluating what they have done.
- **OR** students read through the section headed 'Does light intensity affect the rate of photosynthesis?' to the end of page 171 and complete the questions. In this option, make sure that students have the opportunity to see a sample of pond weed bubbling oxygen in a light source so they understand how the practical works.
- End with a brief plenary summarising the conclusions that can be drawn from the results and the limitations of these conclusions.

9.4 Student Book answers

1.

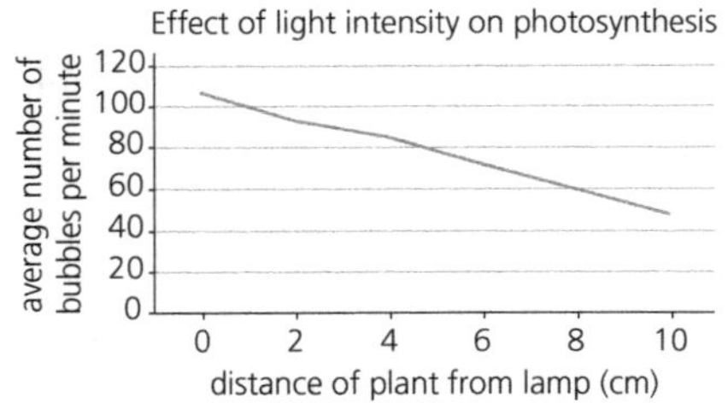

2. The graph shows that the numbers of bubbles produced per minute by the plant decreases as the plant is moved further away from the lamp.

3. That light intensity affects the rate of photosynthesis, and the rate of photosynthesis increases as the light intensity increases.

4. To measure the temperature of the water in the beaker around the test tube containing the experimental pond weed. It is important because temperature might affect the rate of photosynthesis so it is a variable which must be controlled.

5. a. Any two from: Giving the plant time to adjust to the changes in light intensity; measuring the distance from the lamp accurately; controlling the temperature of the water so it stays the same throughout; any other sensible points.

b. Keep the pondweed in the dark before starting the experiment; control the amount of carbon dioxide available to the plant in the water; repeat the investigation more times/with different pieces of pond weed; any other sensible suggestions.

9.5 The need for minerals

Student Book pages 172–173

kerboodle

Online resource

- Worksheet 9.5.1 The mineral needs of plants
- Interactive test

9.5 Student Book answers

1. Water-soluble substances that cells can absorb and need to grow well.
2. Magnesium to make chlorophyll, the green colour needed to capture light energy for photosynthesis.
   Nitrates to make the proteins that control reactions in plant cells and are part of the cell structure.
3. a. Each year the plants take minerals/nitrates from the soil so there is less for the next crop. This year the plants are deficient in nitrates. Know this because symptoms of nitrate deficiency are poor growth and yellowing of the older leaves.
   b. Peas and beans are legumes so they make their own nitrates in root nodules full of bacteria. So they are not deficient in nitrates and they grow well.
   c. Grow other plants where the peas and beans grew this year as they add nitrates to the soil. If students puts compost/ manure/fertiliser, give 1 mark.

Objective

- Plants need minerals to maintain healthy growth and life processes

Overview

This lesson introduces students to the mineral requirements of plants. It is a good opportunity to revisit the mineral requirements of people/animals to help students to understand that all living things face similar problems and often solve them in similar ways. This lesson is an opportunity for a practical exercise which requires at least two-weeks' observation – students can set it up in this lesson and revisit it before the end of the topic **OR** you can set it up at least two weeks before the lesson and demonstrate it on the day.

Activities

- Before the lesson, look back at Unit 6.4 'A balanced diet' and produce several quiz questions to test students' knowledge and understanding of the importance of mineral salts in the human diet.
- Read through page 172 to the end of the section on 'The need for minerals'. Discuss the importance of nitrates and magnesium to plants. Challenge students to recognise the similarity between the need for iron in humans and the need for magnesium in plants. Students answer questions 1 and 2 on page 173.
- Read through 'Mineral deficiencies in plants' with students. Look closely at the images in the table of mineral deficiencies and quiz students on how the symptoms of deficiency diseases in the plants are linked to the minerals they lack.
- Give students **Worksheet 9.5.1**. **EITHER:** Talk through the investigation described. Students set up the practical and revisit their plants every lesson for the next two weeks or more. **OR:** Set up the practical described on **Worksheet 9.5.1** two or three weeks before this lesson. During the lesson, demonstrate the experimental procedure/get students to set up the experimental procedure and then produce plants that have been growing in the different solutions for several weeks, so that students can observe the results and complete the worksheet in this lesson. They can also observe their own investigations over time.
- Work through the final section on page 173 'Where do plants get their minerals from?' with students. If possible, bring a leguminous plant into the lesson to demonstrate the root nodules. Students answer Q3 on page 173.
- Draw the lesson together with a quick review of what has been learned and tell students that they will be using their knowledge and understanding from this lesson in the next one when they look at fertilisers.

Extension

Students research other mineral requirements of plants, e.g. phosphates, potassium.

Homework

Workbook page 72

Key words

minerals, nitrates, magnesium, mineral deficiency, legumes

9.6 The use of fertilisers

SIC

Student Book pages 174–175

kerboodle

Online resources

- Worksheet 9.6.1: The effect of fertilisers on canola and wheat yields
- Worksheet 9.6.2: Food for plants
- Interactive test

Objectives

- Plants need minerals to maintain healthy growth and life processes
- Describe how science is applied across industries

Overview

This lesson helps students to develop their understanding of Science in context.

It gives your students the opportunity to apply their knowledge and understanding of the mineral requirements of plants to the global use of fertilisers to improve crop yields. You also challenge students' recall of their work on decomposers and compost production in Chapter 3 (pages 62–77). Draw on local experience of fertilizer use – which may be mainly artificial, mainly natural or a mixture of both – in your discussions. Help your students make connections between the chemistry they learn in CLSS about non-metals, gases, inert elements and chemical reactions and their biology, by introducing the Haber–Bosch reaction and its importance in making fertilisers from the nitrogen in the air. If students develop the ability to apply what they learn in each science to the other sciences they learn during their lower secondary course, they will have a big advantage when they tackle IGCSE science courses.

Activities

- Students discuss the difference between plants growing wild and farmed crop plants before reading the first two paragraphs of page 174.
- Read through with students the section on page 174 headed 'What are fertilisers'. Make time for questions and discussion about any fertilisers students are familiar with – on local farms, in gardens or for house plants. Take the opportunity to discover whether students remember their work on compost and decomposers from Chapter 3 in Stage 7. Students answer questions 1 and 2 on page 175.
- Read through the rest of pages 174–175 with students. Stop to talk about each paragraph in turn, and make sure that students recognise the huge benefits fertilisers have brought to people. Ask questions such as:
  - Nitrogen is a relatively inert gas – what does this mean?
  - Nitrogen and hydrogen are both non-metals – what does this mean?
  - Nitrogen and hydrogen are gases – what does this tell you about their properties?
- Discuss the value of applying the science of Haber's reaction to the industrial process developed by Bosch, using the data in Figs 9.6 D and 9.6 E (reproduced on **Worksheet 9.6.1**) to support your case. Students answer Q3 on page 175 and then use **Worksheet 9.6.1** to enable them to answer Q4 on page 175 more easily.
- Ask students to complete **Worksheet 9.6.2**. Explain that they have to use their knowledge and understanding of the mineral requirements of plants and the importance of fertilisers and apply it to new information they are given.

Extension

Ask students to investigate the Haber–Bosch process and write a paragraph about how the process works.

Homework

Workbook page 73

Key words

natural fertilisers, manure, artificial fertilisers

9.6 Student Book answers

1.a. A substance that replaces minerals such as nitrates in the soil.

b. Naturally plants take minerals from the soil when they grow, and the minerals are replaced when they die and decompose. Crop plants take minerals from the soil and when we harvest them, no minerals are replaced. Fertilisers are needed to replace the minerals so the crops grow well.

2.a. Advantages: two from, are cheap/easily available/improve the soil structure. Disadvantages: release minerals slowly, limited supply.

b. Advantages: two from, are always available, release minerals fast, farmers control the amount of minerals applied. Disadvantages: expensive, don't improve soil quality.

3. Scientists understood the way plants need minerals, and they developed a method of making ammonia from nitrogen in the air. Industry developed a way to use that reaction to make ammonia on an enormous scale, so artificial fertilisers were available and farmers could increase their crop yields, feeding more people.

4.a. i. No fertiliser → 1300 kg/hectare; with 45 kg fertiliser, yield 2500 kg/hectare → 2500 − 1300 = 1200 increase. 1200/1300 × 100 = 92.3%

ii. No fertiliser → 1300 kg/hectare; with 90 kg fertiliser, yield 3700 kg/hectare → 3700 − 1300 = 2400 increase. 2400/1300 × 100 = 184.6%

b.

kg nitrate fertiliser/ hectare	% increase in yield canola	% increase in yield wheat
45 kg	92.3	60
90 kg	184.6	85

9.7 Water and mineral transport in plants

Student Book pages 176–177

kerboodle

Online resources

- Worksheet 9.7.1 Moving water into a plant: the roots
- Worksheet 9.7.2 Moving water through a plant: xylem tubes
- Worksheet 9.7.3 Losing water from the leaves: stomata
- Interactive test

Objectives

- Describe the pathway of water and mineral salts from the roots to the leaves in flowering plants
- Make conclusions by interpreting results

Overview

In this lesson, you will introduce your students to the basic processes in the movement of water from the soil to the leaves through the plant. This is the basic transpiration stream. Knowledge of the factors affecting transpiration are NOT required in CLSS – that will be covered in IGCSE. In CLSS, focus on the basic plant structures involved in the movement of water through the plant – root hairs, xylem and stomata – and their functions. Give students the opportunity to observe these different parts of the plant if possible. This requires preparation – setting up cress seeds to germinate, placing celery or other stems in water containing ink or food colouring several days before the lesson and painting the underside of leaves with nail varnish. If students develop a clear understanding of the basic principles of the movement of water and mineral salts through a plant, they will grasp more complex aspects of the transpiration stream more easily when they meet it in IGCSE Biology.

Activities

- Ask students why plants need water, and where they get the water from. Ask them about the mineral needs of plants too and how the minerals are transported (Units 9.5 and 9.6).

Prior learning

- Explain observations that plants need water to grow
- Know that water is taken in through the roots and transported through the stem

- Read the first three paragraphs on page 176 with your students, to the end of 'Getting water into the plant'. Ask students to recall the work they did in Unit 2.6 in Stage 7 on specialised plant cells and then give them **Worksheet 9.7.1** to carry out the practical work and complete.
- Read through the section 'Transporting water and minerals around the plant' with your students. **EITHER** give students **Worksheet 9.7.2** to work through **OR** do the sectioning of the stems as a demonstration, showing students the dyed xylem tubes up the stem.
- Read through the section headed 'Leaving the plant' on page 177 before giving students **Worksheet 9.7.3** to carry out the practical. Alternatively, demonstrate looking at the stomata on the varnish peel and project the microscope image for students to see.
- Students work through the Thinking and working scientifically box – ask them why Jamillah's idea of dipping one of the celery stems in wax is so useful.
- Use Fig 9.7 E as a summary with students of the learning in this lesson. Students draw and annotate their own diagram and use any remaining time to answer some or all of questions 1–3 on page 177.

Extension

Students complete questions 1–3 on page 177 and then to think about and answer Q4.

Homework

Workbook page 74

Key words

transpiration stream, stomata

9.7 Student Book answers

1. Roots have root hair cells – microscopic hairs that increase the surface area for water to move in. Soil water moves into the root hair cells by diffusion into the xylem tubes in the root.
2. Mineral salts move into the roots dissolved in the soil water.
3. **a.** Holes found on the underside of leaves through which gas exchange takes place. They can be opened and closed by the guard cells.
   b. Water evaporates from the cells in the leaf and moves out by diffusion through the stomata.
4. Plants lose water through their leaves by transpiration. Water is pulled up through the plant from the soil in the transpiration stream. On cool days, the plant transpires and there is enough water in the soil to replace it. On hot days, more evaporation takes place so more transpiration takes place; there is not enough water in the soil to replace it. The cells of the plant do not get enough water and so they cannot support the plant and it wilts.

Extension 9.8 Xylem, phloem and plant pests

Student Book pages 178–179

kerboodle

Online resources

- Worksheet 9.8.1 Comparing xylem and phloem 1
- Worksheet 9.8.2 Comparing xylem and phloem 2
- Interactive test

Prior learning

- Know that plants have roots, leaves, stems and flowers
- Explain observations that plants need water and light to grow
- Know that water is taken in through the roots and transported through the stem
- Know that plants need healthy roots, leaves and stems to grow well
- Content of Chapter 9 Plant biology

Objective

- Describe the pathway of water and mineral salts from the roots to the leaves in flowering plants

Overview

In this extension lesson, students look in more detail at the transport tissues in plants, learning more about transport in the phloem vessels. This moves their basic studies at CLSS towards the more detailed work on the xylem and phloem that are part of IGCSE Biology. Plant pests and diseases are the context for this lesson. This provides students with an interesting insight into why plant biology is so important. If possible, bring in a plant infected with aphids to show students as part of the lesson.

Activities

- Have a brief discussion with students about the problems of plant pests and plant diseases both locally and globally. Help them to understand both the loss of crop plants and the loss of ecosystems from plant diseases.
- Remind students that they have mainly focused on the transport of water and mineral salts from the soil up through plants in the xylem. Help them to understand that plants need to move the glucose they make in photosynthesis to all parts of the plants to provide the fuel for cellular respiration. Read 'Sugar transport in plants' on page 178 with students, looking at Fig 9.8 A.
- Depending on the abilities of your students and where you feel they need the most practice, give students **EITHER Worksheet 9.8.1** to complete the table, including diagrams of the different tissues **OR Worksheet 9.8.2** to complete a piece of free writing with diagrams describing the structure and functions of the two transport tissues in plants, discussing the similarities and differences between them.
- Introduce the idea that many plant pests attack the transport systems of plants, especially the phloem and ask students why. Read 'Targeting transport' down to the bottom of page 178 and look at Fig 9.8 C. If you have a plant infected with aphids, let students look at it and see how the insects attack the plant.
- Students answer questions 1–4 on page 179.
- Finally, students look at the section headed 'Unwelcome visitors' on page 179. Work through it with students, emphasising the way in which plant pests often introduce plant pathogens into the structure of plants. Make clear: a) how plant diseases that block or damage the transport system of the plant will lead eventually to plant death; and b) the similarities between how these plant diseases and some human diseases are spread by insects penetrating the transport system.

Extension

Students carry out independent research and write a short report comparing the spread of plant diseases by pests such as aphids and human diseases such as malaria spread by mosquitoes.

Homework

Workbook page 75

9.8 Student Book answers

1. Xylem is dead tissue, phloem is alive; xylem transports water and dissolved minerals, phloem transports water and dissolved food/sugars; xylem transports from the roots to the shoots/up the plant only, phloem transports all around the plant, transport in xylem doesn't use energy, transport in the phloem does.
2. **a.** An insect that feeds on the liquid in the phloem of living plants.
   b. Many aphids take a lot of the food from the plant by feeding on the contents of the phloem; the biting mouthparts of the aphids can carry pathogens into the plant, causing diseases.
3. Plants make sugars by photosynthesis and they are carried around the plant in the phloem. They are carried to the buds and provide the food needed to grow many healthy flowers. Aphids stick their stylets into the phloem and feed on the sugary liquid. A plant infected with many aphids will have less sugar reaching the buds, so it will have less food available and so will produce fewer, smaller flowers.
4. **a.** A tree carries out photosynthesis in the leaves. It needs a supply of water carried from the roots to the leaves in the xylem. The sugars made during photosynthesis are carried to all of the tissues of the tree, including the roots, in the phloem. If deer eat a complete ring of bark, both the xylem and the phloem are destroyed. Water cannot reach the leaves so they cannot photosynthesise and die. Sugars cannot reach the roots so the cells are starved and the roots die. These two things mean the whole tree dies.
   b. Covering the bark of young trees above the level that can be reached by deer, etc. – until the tree is older and the bark is too tough for the deer to eat it.

9.9 Review answers

Student Book pages 134–135

Student Book answers

1. **a.** Photosynthesis is the process by which plants make their own biomass/food using carbon dioxide and water and light energy captured by chlorophyll. **[2]**
   b.

   $$\text{carbon dioxide + water} \xrightarrow[\text{chlorophyll}]{\text{light}} \text{glucose + oxygen}$$

   (reactants) (products) **[3]**

   c. Carbon dioxide gets into the leaf from respiration in the cells and diffusion from the air through the stomata.
   Water enters the root hair cells by diffusion from soil water and is transported up the plant in the xylem to the photosynthesising cells in the leaves.
   Oxygen is used by the plant cells for aerobic respiration and the rest is lost through the stomata in the leaves by diffusion.
   Some of the glucose is used directly by the cells for aerobic respiration, some is converted to starch to store energy, some is used to build other molecules such as proteins, and some is transported around the plant to all of the other cells to use. **[10]**
   d. **i.** No – root hair cells have no chloroplasts and they are not exposed to light. **[3]**
   ii. Yes – the light needed for photosynthesis is captured by chlorophyll in the chloroplasts when there is plenty of light. **[2]**
   iii. No – chloroplasts are still there but there is no light for them to capture. **[2]**
2. **a.** The pond weed photosynthesised in the light. It absorbed carbon dioxide from the water and used it in photosynthesis. The concentration of carbon dioxide in the water fell so the indicator turned purple. **[3]**
   b. The pondweed did not photosynthesise in the dark, so it did not remove any carbon dioxide from the water. It respires all the time, in the light and the dark, so in the dark it added carbon dioxide to the water from respiration and the indicator changed colour. **[4]**

3. a. He collects the gas produced by the plant in a measuring cylinder. **[1]**

b. **[5]**

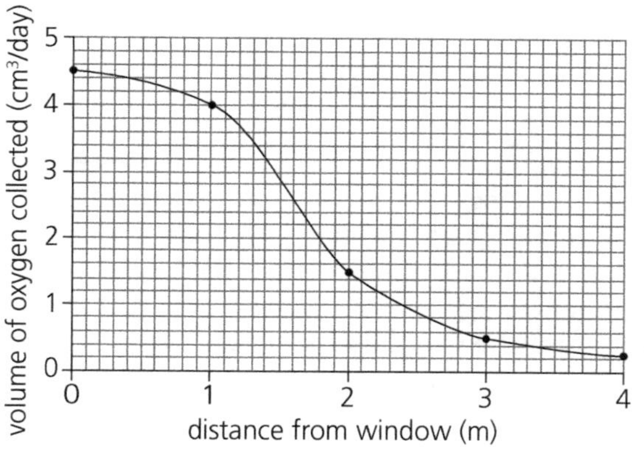

c. The closer the plant is to the light of the window, the more oxygen is produced – indicating that more photosynthesis is happening. So the greater the light intensity, the more photosynthesis takes place. **[3]**

d. Any two sensible suggestions, e.g. temperature of water, amount of carbon dioxide in the water, the size of the piece of pondweed. **[2]**

4. a. Does a plant need carbon dioxide to photosynthesise? **[1]**

b. To use up any starch it has stored in its leaves, so the presence of starch can be used to indicate photosynthesis has taken place in the investigation. **[2]**

c.
- Take the two experimental leaves from the plant and drop them into boiling water to remove the waterproof layer and break open the cells.
- Turn off the heat. Take the leaves from the boiling water.
- Place the leaves in a test tube of ethanol and put the test tube into the hot water. The ethanol will boil and the green colour will come out of the leaves.
- Remove the white leaves from the ethanol and dip them into hot water to soften them.
- Spread the leaves on a white tile and add a few drops of iodine solution to each. **[6]**

d. Leaf B (from the bag containing lots of carbon dioxide) will turn the iodine blue–black. It has been photosynthesising and contains starch. Leaf A (from the bag with no carbon dioxide) cannot photosynthesise so it has no starch in its leaves and has no effect on the iodine which stays yellow–brown. **[5]**

5. a. The amount of minerals available to the plants. **[1]**

b. Minerals are absorbed into the roots of plants from the soil, dissolved in the soil water. They get into the plant through the root hair cells. They move into the xylem and are transported up the plant to all the cells in the transpiration stream. **[4]**

c. i. Magnesium is needed to make chlorophyll which is the green colour in plants. If a plant lacks magnesium, it can't make chlorophyll and this affects the colour of the leaves **[2]**

ii. Chlorophyll captures light energy for photosynthesis. If a plant lacks magnesium it cannot make the chlorophyll it needs, so it cannot capture as much light energy. Less photosynthesis takes place, so less food is made and so the plants do not grow as quickly as those that have plenty of magnesium. **[5]**

6. a. Xylem **[1]**

b. It is a dead tissue comprising tubes that go from the roots to the leaves and buds; it carries water (and dissolved minerals) up the plant from the roots to the leaves. **[3]**

c. It was put in plain water/it had the end sealed up, e.g. by wax, before it was put in the ink. **[2]**

10.1 What is excretion

Student Book pages 182–183

kerboodle

Online resources

- Worksheet 10.1.1 Excretion wordsearch 1
- Worksheet 10.1.2 Excretion wordsearch 2
- Interactive test

Prior learning

- Use the scientific names for some major body organs
- Describe the main functions of major body organs

10.1 Student Book answers

1. Excretion means getting rid of/removing waste products made in the body.
2. Removing undigested food from the body.
3. Kidneys
4. Urea comes from the breakdown of excess proteins in the liver. It is toxic so it must be excreted so that it does not poison/cause damage to the organs of the body, e.g. the brain.
5. The kidneys do a very important job removing toxic urea, so it is important that they do not get damaged. The thick layer of fat which covers them is to protect them from damage.

Objective

- Describe the structure of the human excretory system

Overview

This is the first lesson on excretion for students. A common misconception in students, even at IGCSE level biology, is that defaecation is excretion. This initial lesson is a good opportunity for you to make sure students that know and understand that excretion is the removal of waste products made in the cells of your body, why it is so important, and the main excretory organs of the body. THEN you can explain clearly why the removal of undigested food and dead cells in the faeces is NOT excretion but egestion.

Activities

- Remind students that excretion is one of the characteristics of all living organism. Give a quick quiz to see what your students understand about excretion at this stage. Ask questions such as 'What are your kidneys?', 'Where are the kidneys in your body?', 'What do your kidneys do?', 'What do we mean by excretion', 'What is excreted in your lungs?'.
- Read through the first two paragraphs on page 182. Make sure students understand that waste products such as urea and carbon dioxide are toxic.
- Read on through 'What do we excrete?' Remind students that they looked in detail at the removal of carbon dioxide by the lungs in Stage 8 Chapter 5, and so in this chapter they will be focusing on the excretion of urea by the kidneys. In their work on balanced diets in Stage 8 Chapter 6, they learned how excess carbohydrates and lipids in the diet are converted to fat and may contribute to obesity. Explain that the body has to breakdown excess protein that is eaten. Part of the molecule is removed and forms toxic urea in the liver which must be removed, and this is one of the main functions of your kidneys.
- Read 'What do kidneys look like?' and look at Fig 10.1 C. Emphasise the importance of the protective fat around the kidneys. If possible, bring in a kidney from a butcher and dissect it for them, removing the fat, identifying the tubes that carry urine out of the kidney and the blood vessels, and then cutting the kidney in half to show the different layers of tissues.
- Have some urea crystals and water – show students that urea crystals are white and dissolve in water to make a colourless solution. Explain that the different colours of urine are a result of other compounds that are excreted and depend on the concentration of the urine.
- Discuss the difference between excretion and egestion with students before they read the final paragraph headed 'Excretion or egestion?
- Students answer questions 1–5 on page 183.
- Give students either **Worksheet 10.1.1** or **Worksheet 10.1.2**. **Worksheet 10.1.1** is a wordsearch using key words involved in excretion, which helps to give weaker students or students who lack confidence with scientific language an opportunity to focus on and learn these key terms. **Worksheet 10.1.2** tests students' knowledge and understanding of the basic principles of excretion as well as building familiarity with important terminology. They complete a series of sentences before finding the words they use in their answers in the wordsearch.

Extension

Students make a table to compare excretion and egestion.

Homework

Workbook page 76

Key words

toxic, kidneys, urine, faeces, egestion

10.2 The human excretory system

Student Book pages 184–185

kerboodle

Online resources

- Worksheet 10.2.1 The human excretory (renal) system
- Worksheet 10.2.2 Extension: Balancing the blood
- Interactive test

Prior learning

- Use the scientific names for some major body organs
- Describe the main functions of major body organs

Extension

Give students **Worksheet 10.2.2**. Students have to use their knowledge and understanding of the function of the kidneys to predict urine production in two different situations. They have to explain the graphs they draw.

Homework

Workbook page 77

Key words

ureter, bladder, urethra

Objective

- Describe the structure of the human excretory system and its function

Overview

This lesson continues directly from Unit 10.1. Students move on from considering excretion and the basic appearance of the kidney to learning the structures and functions of the human renal excretory system. By the end of lesson, they should know the basic structure of the human renal system, the function of the kidney as a filter and the importance of the kidney in both removing toxic urea from the blood and maintaining the water balance of the body. If they develop a basic understanding of these processes now, they can build on it when they study excretion in IGCSE Biology.

Activities

- Ask students to summarise the key points from the previous lesson – write their suggestions on the board.
- Explain to students that the kidneys are part of a complex excretory system. Read through the sections 'How do we remove urea?' and 'Making and storing urine' with students. Give them the opportunity to ask questions. A useful demonstration is to remind students about the principles of filtration with a simple filter funnel, filter paper, and a mixture of sand and salt which you mix with water and then filter.
- Give students **Worksheet 10.2.1**. Students should label each structure and annotate the diagram by adding the functions of the different parts of the system.
- Students answer questions 1–4 on page 185.
- Ask students to estimate how much urine is produced by the average human kidneys in a day. (Depending on your students, keep this very general rather than personal). There may be quite a range! Remind students that the concentration of their blood must remain the same within narrow limits, and explain that the kidneys are very important in keeping this concentration relatively constant. Ask them to suggest things that might affect the water balance of the body (e.g. amount of drink, exercise done, temperature/humidity of the air, amount of sweating, state of health).
- Students complete Thinking and working scientifically Q5 using the data in Fig 10.2 C.
- Have a plenary session summarising what students have learned.

10.2 Student Book answers

1. Urine is the yellow liquid produced by your kidneys – it contains urea, water and other chemicals.
2. Your kidneys filter the blood – urea passes through the filters but blood cells and big, useful molecules stay in the blood. Small useful molecules may pass through the filter but they are taken back into the blood.
3. The ureter carries urine from the kidney to the bladder; the urethra carries urine from the bladder to the outside world.
4. Similarity: both contain urine. Difference: one from, kidneys make urine by filtering the blood; the bladder stores urine made by the kidneys.
5. After you have drunk a lot of water, the amount of urine produced by the kidneys increases over time until it reaches a peak. As the extra water is removed, the volume of urine produced falls back towards normal.

10.3

Who made the best model?

TWS

Student Book pages 186–187

kerboodle

Online resource

- Worksheet 10.3.1 Making a model kidney
- Interactive test

Extension

Students could make the model they have planned.

Homework

Complete planning their model kidney or answer Workbook page 78.

Objectives

- Describe some important models and discuss their strengths and limitations

Overview

This lesson helps students develop to their skills in Thinking and working scientifically. Students are presented with two different models of the kidney and asked to evaluate them, discussing their strengths and weaknesses. Analytical thinking and an understanding of the value and limitations of models in science are valuable skills for students to develop, and will help them to excel when they continue studying Biology at IGCSE.

Activities

- Begin the lesson by pointing out that some students find the kidney difficult to understand and reminding students that scientists use physical models to help explain complex systems.
- **EITHER** read through this spread with students **OR** ask students to read about the two different models themselves.
- Have a class discussion about which model your students find most helpful. Ask them to suggest strengths and weaknesses of the different models and have a class vote on which the students prefer. Compare the results of your class to the results given in Fig 10.3 C.
- Students complete questions 1–5 on page 187.
- Once students have completed the questions, give them **Worksheet 10.3.1**. Each student plans a model kidney – they may be very similar to those described in the Student Book or quite different.

10.3 Student Book answers

1. A physical model is something used by scientists to help explain something which is too small, too big or too difficult to see, such as the filtering system in the kidney.
2. <u>Strengths</u> e.g. it shows filtering in action/you can see how some things come through the filter and others don't/ you can see that the filtering depends on the size of the particles/it has substances in solution like the real thing/any sensible point.

   <u>Weaknesses</u> e.g. it looks as if all the water is filtered out of the blood through the model kidney– this doesn't happen/ looks as if some of the red blood passes through the filter which it doesn't/any other sensible point.
3. Strengths e.g. it shows how the kidney filter acts as a sieve/you can see how small particles come through the filter and larger ones don't/ you can see that the filtering depends on the size of the particles/any sensible point.

   <u>Weaknesses</u> e.g. There is nothing that is actually in solution so it is unrealistic/it looks as if all the water is filtered out through the sieve/ any other sensible point.
4. For example, Ejaz's model: don't add the red colouring because blood does not go through the filters in the kidney/Fatima's model: reduce the size of some of the holes so some of the water does not pass through the filter/any sensible ideas.
5. $22/40 \times 100 = 55\%$

10.4 When kidneys go wrong

SIC

Student Book pages 188–189

kerboodle

Online resource

- Worksheet 10.4.1 Kidney dialysis
- Interactive test

10.4 Student Book answers

1. Because urea is toxic and the body cannot get rid of urea, excess water and other waste products if the kidneys do not work. This affects all the other body systems and makes people very ill.
2. Reasons for being better or worse include levels of infectious kidney disease, levels of diabetes, available health care, etc.
3. Dialysis: instant results, makes patients feel better and carry on with relatively normal life BUT has to be repeated often, takes a lot of time, expensive, not enough machines, eventually damages the blood, any other valid points.

   Transplants – permanent, relatively cheap, allows patient to live a normal life going forward BUT ethical decisions to be made about donors, e.g. should live donors be allowed to give away a kidney? Some people worry about the use of organs from deceased donors, etc.

Objectives

- Evaluate issues which involve and/or require scientific understanding

Overview

This lesson helps students to develop their understanding of Science in context. It presents a good example of some of the ways in which science affects us in our everyday lives. It shows how scientific understanding enables us to treat diseases that would otherwise result in the loss of life. It also reinforces understanding that science cannot answer everything – that moral and ethical considerations also affect the choices we make. It is useful to check out the laws and attitudes of your country regarding organ donation before using this lesson. In some countries, individuals must opt in to organ donation schemes; in others they must opt out if they do not want to donate. In some countries, organs are paid for. In others they are freely given.

As always when dealing with medical issues, be very sensitive in dealing with this topic as some students or their family members may be affected by kidney failure.

Activities

- Read the first paragraph with students and ask them to suggest why someone feels ill if their kidneys are not working properly. Use this to confirm students' understand the role of the kidneys in the body.
- Read the section 'Life without kidneys' to give students an understanding of the size of the problem. Many students may be unaware of the problems caused by kidney failure or that 4.6% of all deaths are due to kidney failure.
- Students answer questions 1 and 2 on page 189.
- Read 'Life without kidneys: a kidney machine' with your students. Discuss the advantages and disadvantages of using kidney dialysis. Ask students why Mohammed feels well after each dialysis session but feels ill again before the next one. Tell students that Mohammed has to control what he eats and drinks most of the time, but eats and drinks whatever he likes for the first hour or two he is on dialysis and ask them to suggest reasons why.
- Give students **Worksheet 10.4.1** They list the substances removed from the blood in the dialysis machine and the substances that remain and make a table to compare a dialysis machine with working kidneys.
- Read through 'A new kidney for life', looking at kidney transplants from both living and deceased donors. Describe the problems of tissue matching and rejection and explain that, unless a transplant comes from an identical twin, anyone who receives a new kidney has to take medicine for the rest of their life to make sure their body does not attack the new organ.
- Give students the opportunity to discuss the issues which might arise around organ donation before they answer Q3 on page 189.

Extension

Students research the ways in which different countries organise their organ donation programmes.

Homework

Workbook page 79

Key words

dialysis, kidney transplant, donor

Extension 10.5 Kidneys work everywhere!

Student book pages 190–191

kerboodle

Online resource

- Worksheet 10.5.1 Urine concentration of different mammals
- Interactive test

Extension

Write about the importance of kidneys in vertebrate survival across the world.

Homework

Workbook page 80

Objective

- Describe the structure of the human excretory (renal) system and its function

Overview

In this Extension unit, students apply what they have learned about human kidney structure and function to the renal excretory systems of animals living in a variety of challenging habitats. They need to demonstrate clear understanding of the demands of the different habitats described and the need of animals to maintain water balance in their tissues and also to remove the urea they produce. This moves students towards the work on adaptive features that they will meet in IGCSE Biology, and also introduces the importance of homeostasis, another important aspect of IGCSE Biology.

Activities

- Ask students to list some of the very different places where people live around the world. Ask them how we manage to survive, to keep our blood concentration constant and remove the urea we produce regardless of where we live. Draw out the idea that humans change their environment – storing water, heating or cooling homes, changing clothes, etc. and this enables us to live on land across the globe.
- Read through the section on page 190 'Hot and dry' with students. Discuss ways in which animals are adapted to hot, dry conditions – see Figs 10.5 A and 10.5 B. Students answer questions 1 and 2 on page 190.
- Read through the section on pages 190–191 'Water everywhere' with students. Discuss different challenges of fresh water and salt water environments and some of the methods of coping – reference Figs 10.5 C and 10.5 D. Students answer Q4 on page 190.
- **EITHER** ask students to complete page 191 Q3 **OR Worksheet 10.5.1** which gives them the data and squared paper to complete the bar chart.
- Ask students to work individually or in groups to research the excretory systems of other vertebrate groups: birds, reptiles, amphibians and fish. They should find out the organs involved, the main toxic substance excreted, and whether the organisms are able to produce urine more concentrated than their blood.

10.5 Student Book answers

1. Camels are larger animals than humans but they produce less urine. Average human urine 1500 cm^3, average camel urine 500–1000 m^3.
2. Camels are large, kangaroo rats are small; camels need to drink, kangaroo rats do not need to drink as they can get enough water from the breakdown of their food; both have specially adapted kidneys which can produce very concentrated urine, but kangaroo rats can produce more concentrated urine than camels.
3. a.

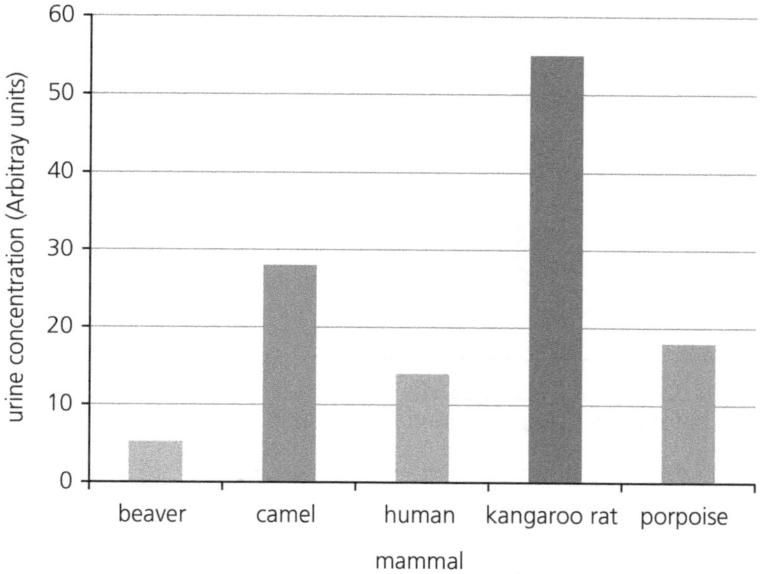

   b. Advantage: It shows very clearly which animals produce the most concentrated urine. Disadvantage: some people might misread it and think it showed which animals produced the biggest volume of urine.
4. Kangaroo rats have little or no water available to them. Often the only water they get is from the breakdown of their food. They produce very small amounts of highly concentrated urine – approximately 4× as concentrated as human urine. Porpoises have limitless water available but it is salty. They have to get rid of the excess salt which uses a lot of water, so they produce large amounts of relatively concentrated urine – 1.3× the concentration of human urine.

10.6 Review answers

Student Book pages 192–193

Student Book answers

1. a. Excretion means getting rid of/removing waste products made in the body. [2]
   b. Urea, carbon dioxide [2]
   c. Waste products like urea and carbon dioxide are toxic/poisonous so if they build up, they damage the tissues and cause illness or death. [3]
2. a. A toxic waste product formed from the breakdown of protein in the body. [2]
   b. The liver c. The kidneys [1, 1]
   d. They do a very important job so it is important to protect them from damage. This is what the fat is for. [2]
   e. Poisonous/damages the body [1]
3. **Urea** is very toxic. If the concentration in your **blood** gets too high, it damages other organs including your brain. It can make you very ill and may lead to **death**. It is removed from your body by special organs called the **kidneys**. They **filter** the toxic molecules from your **blood** and pass it out of your body as part of your **urine**. [7]
4. a. Excretory system b. Renal system OR **a.** and **b.** given vice versa [1, 1]
   c. To remove toxic urea and excess water from the blood. [2]
   d.

Part of the system	Function
Kidney	Filters the blood and removes urea, excess water and other substances not needed by the body
Ureter	Carries urine from the kidney to the bladder
Bladder	Stores urine
Urethra	Carries urine from the bladder to the outside world

5. a. Bar chart with volume of urine in cm^3 on vertical axis and time intervals 30 mins, 60 mins, etc. along the bottom. [7]

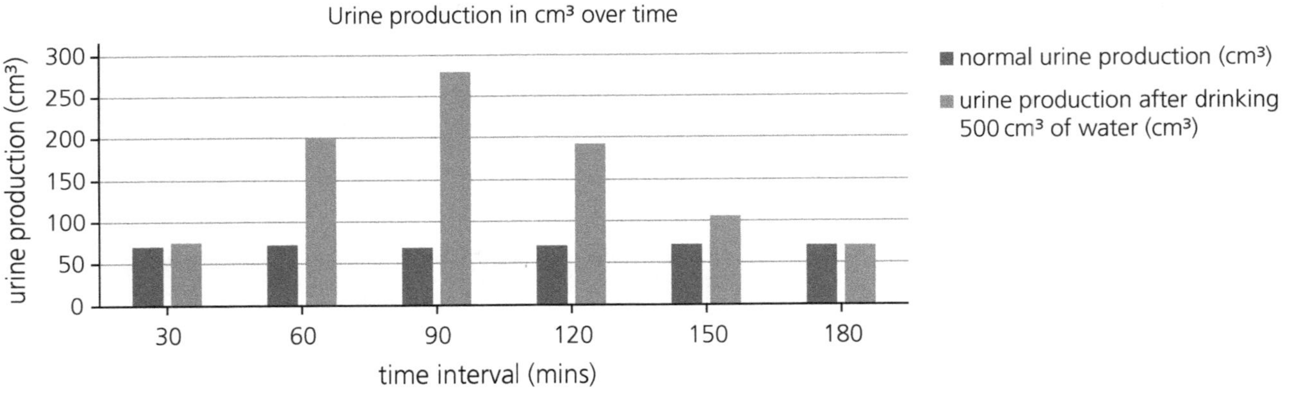

   b. Normal volume of urine in 30 mins is 75 – 70 = 5 cm^3; 200 – 72 = 128 cm^3; 280 – 68 = 212 cm^3; 190 – 70 = 120 cm^3; 105 – 71 = 34 cm^3. The patient produced 5 + 128 + 212 + 120 + 34 = 499 cm^3 of extra urine. [7]
6. a. Your kidneys filter your blood. Toxic waste products such as urea pass through the filters and are removed from your blood. Blood cells and big molecules which your body needs stay in the blood.
   Some useful substances, like glucose, are also small enough to pass through the filters in the kidneys. They get taken back into your blood because your body needs them. The yellow liquid formed is called urine. [5]
   b. Some water is filtered out of your blood, along with other substances which your body does not need, e.g. excess salt from your food. [2]
   c. If the kidneys fail, then urea is not removed from the body. Urea is toxic so, if the levels build up, it can damage other organs and cause death. If the levels of salt and water in the blood are wrong, this can cause problems. [3]
   d. It can't be with you all the time/it is expensive/it is not as good at balancing the blood as kidneys/toxic substances build up between times on the machine. [3]

11.1 Reproduction: a characteristic of life

Student Book pages 194–195

kerboodle

Online resources

- Worksheet 11.1.1 Extracting DNA
- Worksheet 11.1.2 Asexual reproduction in plants
- Interactive test

Prior learning

- Know that animals, including humans, and plants reproduce

Extension

Students complete Q5 on page 195 of the Student Book.

Homework

Workbook page 81

Key words

genetic material, DNA, chromosomes, genes, asexual reproduction, sexual reproduction, gametes

Objectives

- Reproduction is one of the seven characteristics of living things
- Know that chromosomes contain genes made of DNA

Overview

In this lesson, you introduce students to the importance of genetics in reproduction. They learn the key terms DNA, chromosomes and genes and the part they play in cells. You have the opportunity to provide a stunning demonstration or highly memorable practical for students by extracting DNA from plant material so that they can see it for themselves! Students discover the difference between asexual and sexual reproduction. Source a plant that reproduces asexually such as a spider plant, mother-of-thousands, burro's tail or strawberry plant, and give your students the opportunity to grow their own individual asexually produced plants. This lesson is a foundation for the rest of this chapter and for Chapter 13.

Activities

- To introduce this chapter, remind students about reproduction as a characteristic of all living organisms. Ask them to write down what they think the terms genetic material, DNA and chromosomes mean – most students will have heard of DNA but 'don't know' is an acceptable answer if they have no idea what it is!
- Read through the first three paragraphs on page 194. Look at Fig 11.1 B showing chromosomes in cell nuclei. Remind students of their work on cells and the fact that cells have nuclei. Students do NOT need to know what DNA stands for – simply that it is the genetic material that carries the instructions for new cells, and that it forms the chromosomes as a cell divides.
- Give students **Worksheet 11.1.1**. If possible **EITHER** carry this out as a demonstration **OR** allow students to carry out the practical themselves – whichever is an exciting and memorable experience and helps fix DNA firmly in student minds.
- Now read through the section headed 'Asexual reproduction' with students, looking at Fig 11.1 C. Give students **Worksheet 11.1.2** and let them carry out the procedure, giving them each an identical plantlet to take care of. You can use these later in Chapter 13 when looking at variation.
- Read through the final two sections on page 195 with students: 'Sexual reproduction' and 'What are gametes?' Use Fig 11.1 A to show students that the young geese are not identical to each other.
- Draw together all the main points from the lesson. Students answer questions 1–4 on page 195.

11.1 Student book answers

1. The process by which living things make more of themselves.

2. a. An enormous molecule which is the genetic material of cells. It contains the instructions for making a new life.

b. Chromosomes

3. a. 46

b. 23

4. Asexual: one parent/parent cell/ offspring identical to parent/same DNA.

Sexual: two parents/parent cells/offspring get DNA from both parents/offspring differ from both parents.

5. a. $82/2 = 41$

b. $21 \times 2 = 42$

11.2

Fertilisation: a new life begins

Student Book pages 196–197

kerboodle

Online resources

- Worksheet 11.2.1 Human gametes
- Worksheet 11.2.2 Discovering gametes: the work of Spallanzani
- Interactive test

Prior learning

- Know that animals reproduce

Extension

Students plan physical models of an egg and sperm cell, to show as many features as possible – ideally, the models should demonstrate the process of fertilisation. Students should annotate their plans both to show how to make the model and how it would work.

Homework

Workbook page 82

Key words

eggs, ovaries, sperm, testes, fertilisation, embryo

Objectives

- Know that chromosomes contain genes, made of DNA
- Know that genes contribute to the characteristics of an organism
- Describe the fusion of gametes to produce a fertilised egg with a new combination of DNA

Overview

This lesson continues directly from Unit 11.1. Students should grasp the key facts about male and female gametes, including the fact that each gamete contains half the number of chromosomes as the normal body cells. This will help them to understand sex inheritance in CLSS and the genetics they will study in IGCSE Biology much more easily. You may explain how the sperm are introduced into the female, if this is appropriate for your students.

Activities

- Begin by firing a few questions at students, e.g. garlic cells have 8 pairs or 16 chromosomes – how many chromosomes will there be in the gametes? Elephant gametes each contain 28 chromosomes – how many chromosomes are there in an elephant body cell? How many pairs of chromosomes are there in an elephant body cell? Make sure students know that the gametes contain half the number of chromosomes as the body cells, and that the chromosomes in body cells come in pairs, one from the mother and one from the father.
- Read through page 196 with students and give them **Worksheet 11.2.1**. They label and annotate the diagrams to show how the different gametes are adapted to their functions. **OR** students can answer Q1 on page 197.
- Talk through the process of fertilisation with students, explaining how the egg and sperm fuse to form an embryo with a new combination of DNA. Then ask students to read page 197. They should copy Fig 11.2 C and answer Q2 on page 197.
- Tell students that although we now understand the process of fertilisation, it was unknown for many centuries. Give students **Worksheet 11.2.2** to work through either individually or in pairs. Bring the whole class together and go through the correct answers to the table – use a show of hands to see what students have answered and explain the correct conclusions as you go through. Students answer Q3 on page 197. Use this as the focus for a brief plenary session to make sure that students have grasped the key points of the lesson.

11.2 Student Book answers

1.

Eggs	Sperm
Relatively big	Small
Contain a store of food	No food store
Few	Many
Made in the ovaries	Made in the testes
Can't move themselves	Swim to the egg
No tail	Long tail

2. Many thousands of sperm swim to the egg → one sperm gets into the egg cell → the egg stops any more sperm cells getting in → the nucleus of the successful sperm fuses with the nucleus of the egg cell → the fertilised egg begins to divide to form an embryo.

3. a. Gametes have half the number of chromosomes as normal body cells.

b. At fertilisation, the nuclei of the two gametes fuse – so the new cell has twice the number of chromosomes as each gamete – which is the number of a normal cell.

11.3

Boy or girl? Sex inheritance in humans

Student Book pages 198–199

kerboodle

Online resources

- Worksheet 11.3.1 Human chromosomes
- Worksheet 11.3.2 Sex inheritance in humans
- Worksheet 11.3.3: Male : female ratios
- Interactive test

Extension

Students complete **Worksheet 11.3.3** and carry out independent research, as required.

Homework

Workbook page 83

Objectives

- Describe the fusion of the gametes to produce a fertilised egg with a new combination of DNA
- Sex in humans is inherited through the X and Y chromosomes

Overview

In this lesson, students learn how sex is inherited in humans, which has relevance both in their biology studies and in life. There can be pressure to produce a particular sex – it is valuable to learn that every pregnancy has a similar chance of producing a boy or a girl, and that it is the father's sperm that determines the sex of the embryo.

Activities

- Ask students how many chromosomes there are in a human body cell, how many pairs of chromosomes there are in a human body cell and how many chromosomes there are in human eggs and sperm (46, 23 and 23).
- Summarise the process of fertilisation from the previous lesson using Fig 11.3 A and remind students that half of the DNA in an embryo comes from the mother and half from the father.
- Students read page 198 down to 'The sex chromosomes' and answer questions 1 and 2 on page 198.
- Read through the rest of page 198 and the first two bullet points on page 199 with students. Describe how the two chromosomes in each pair are identical apart from the sex chromosomes. Give students **Worksheet 11.3.1**, and point out the alternative forms of the sex chromosomes.
- Read through the rest of page 199. Make sure students understand that there is a 1 in 2 chance that each embryo will be male or female. Give out **Worksheet 11.3.2** and provide each student with **EITHER** four marbles, counters or similar to represent the chromosomes **OR** four identical small pieces of card. Each student needs three items of one colour to represent X chromosomes and one of a different colour to represent the Y chromosome, or writes an X on three pieces of card and Y on the fourth. Give them time to complete ten 'fertilisation events' as described on the worksheet and collect the data from each student to see how many boys and girls have been produced. This allows you to illustrate two important points: Any particular student may have any number of boys and girls, from all girls to all boys. The bigger the data set, as you collect information from each student, the nearer the class average will come to a 1 : 1 ratio of boys to girls.
- Students answer questions 3 and 4 on page 199 to complete the lesson.

11.3 Student Book answers

1. Cell – nucleus – chromosome – DNA – gene
2. You inherit two sets of chromosomes, one from your mother, one from your father. They make pairs of chromosomes. Genes on the paired chromosomes control the same features. The gene that controls a feature may be from your mother or father.
3. **a.** 23 chromosome pairs

   b. 22 chromosome pairs are exactly the same size. The sex chromosomes may be exactly the same – XX (female) or one may be much smaller than the other – XY (male).
4. Each child is produced when an egg and a sperm fuse. Each egg contains an X chromosome. A sperm may contain an X chromosome or a Y chromosome. There is a 50 : 50 chance that the egg will be fertilised by an X or a Y sperm to make a girl or a boy. This is the same for every child – the sex of any other children does not affect this. (Diagram acceptable).

11.4 Variation between individuals

Student Book pages 200–201

kerboodle

Online resources

- Worksheet 11.4.1 The effect of our environment
- Worksheet 11.4.2 Why are brothers and sisters different?
- Worksheet 11.4.3 Factors affecting height
- Interactive test

Objectives

- Describe genetic differences between individuals
- Know that genes contribute to the determination of an organism's characteristics

Overview

Variation is a key concept in many areas of biology, from adaptation and ecology to natural selection. This lesson enables you to build a solid foundation of understanding about the difference between inherited variation, which is a result of the chromosomes inherited from the parents, and environmental variation. If students understand this, and the way genetic and environmental variation act together in many cases, they will progress rapidly in these areas in IGCSE Biology.

There is also an opportunity in this lesson for students to develop their skills in Thinking and working scientifically by manipulating data.

Activities

- Ask students for examples of characteristics they share with parents, siblings or cousins.
- Read through page 200 with students and discuss the impact of genetic and environmental variation. Ask students for examples. Give students **Worksheet 11.4.1** and work through it with them, supporting them as they consider the impact of the different environmental factors on the individuals described. Encourage them to think about other factors that might also be having an effect.
- Summarise the fact that everyone inherits genetic information from their parents, but the environment each child grows up in is different, even in the same home (having siblings, changing family circumstances, etc). Give students **Worksheet 11.4.2** to consider and complete. Go through the worksheet as a class – highlight the errors and ask some of the students to read out their comments.
- Ask students to read the paragraph 'Identical twins' on page 201 and then answer questions 1 and 2. If you have identical twins in your class, you may find that they are happy to discuss their similarities and environmental differences, but this must be handled carefully as some students would not be comfortable with this.
- Introduce students to the interaction of genes and environment in determining height. Read through the Thinking and working scientifically box and then give students **Worksheet 11.4.3** to complete. This provides them with Table 11.4.1 from the Student Book, along with graph paper for their bar chart and two tasks.

Extension

Students complete Q3b and 3c on page 201.

Homework

Complete **Worksheet 11.4.3** or Workbook page 84.

Key words

inherited variation, environmental variation, identical twins

11.4 Student Book answers

1. Each child has a random selection of genes from each parent, so they will have some in common and some that are different.
2. They are genetically identical and so all of the features controlled by their genes will be the same. The environment will not have had time to make many changes. The older twins get, the more different environmental factors will have an impact on some of their characteristics, e.g. height and body mass are affected by eating habits and exercise, etc. Features entirely controlled by genes remain the same.

3.a.

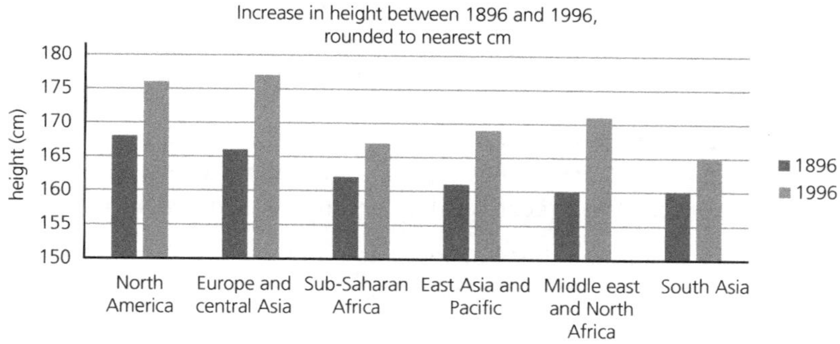

b. i. Percentage increase = (1996 reading −1896 reading)/1896 reading × 100 SA = 3.1%, MENA = 6.9%, EAP = 5.0%, SSA = 3.1%, ECA = 6.6%, NA = 4.8%. Biggest increase is the Middle East and North Africa; smallest is both South Asia and East Asia and Pacific.

ii. Possible reasons: The countries with the biggest increases may have seen the biggest improvements in living standards/food availability, countries with small increases relatively little improvement; countries with relatively little improvement may have had a higher standard of living in 1896 and so been nearer to their genetic height potential, so less scope for getting taller; any other sensible suggestions.

c. i. 1996 reading – 1896 reading for tallest area: (168 – 160) cm so difference in height = 8 cm
1996 reading – 1896 reading for shortest area: (177 – 165) cm so difference in height = 12 cm

ii. In 1896 neither population reached full genetic height potential as both had relatively poor levels of nutrition. NA taller as genetically taller population. By 1996 NA population reaching full genetic height potential as plenty of food available. Bigger difference between tallest and shortest – full genetic height difference showing. It might be that shortest population are genetically much shorter OR, although food levels have improved everywhere, lack of food may still mean full genetic potential not reached in shortest population.

11.5 The development of a fetus

Student Book, pages 202–203

Objective

- Describe healthy fetal development

Overview

This lesson takes students from reproduction at the level of chromosomes and gametes to the development of a fetus in the uterus. This is important both as part of their CLSS course, as preparation for considering factors that have an adverse effect on fetal development in Units 11.6–11.8 and for them as biologically literate citizens understanding their own bodies.

Be very sensitive dealing with this topic as students may have family members who are pregnant, trying to become pregnant, have just had or have even lost a baby.

kerboodle

Online resource

- Worksheet 11.5.1 A growing fetus
- Interactive test

Prior learning

- Know that humans reproduce

Activities

- Ask students how many cells make up a human being – they may remember billions from Stage 7 work on cells. Read through the text on page 202 of the Student Book, down to the heading 'Protected, warm and fed'. Discuss the cell division that follows fertilisation and the movement of the ball of cells into the uterus to continue development. Look at Fig 11.5 A with students and get them to use their rulers to measure the size of the embryo at the stages shown. Students answer Q1 on page 203.
- Point out that the embryo needs increasing amounts of food and oxygen as it grows and its cells produce more waste products. How does it acquire/get rid of these? Project Fig 11.5 B on the whiteboard or display it in some other way. Read through the section headed 'Protected, warm and fed' with students, referencing Fig 11.5 B and pointing out the structures as they are described. **Either** give students **Worksheet 11.5.1** to work through and complete as a record of this content **OR** students answer Q4 on page 203.
- Read through the final paragraph on page 203 describing how a baby is born. Be prepared to answer questions from students on this process and emphasise the role of the mother and the father AFTER birth in caring for the baby and helping it grow to be a healthy adult.
- Students answer Q3, page 203 – ask them to illustrate their flow diagram with drawings of the fetus at different stages to help them to remember it.
- Have a plenary session quizzing students to make sure they have understood the key elements of this lesson.

Extension

Ask students to carry out some research into the structure of the placenta. They should write an illustrated piece about the placenta, explaining how substances pass from mother to fetus and from fetus to mother, and why it is so important that the blood of the mother and the fetus do not mix.

Homework

Workbook page 85

Key words

uterus, gestation period, pregnancy, fetus, placenta, umbilical cord, fluid sac

11.5 Student Book answers

1. **a.** Some of the cells specialise to form particular tissues and organs.
   b. 4 weeks
2. It supports the baby as it grows and moves. It acts as a shock absorber to protect the developing baby from bumps and knocks.
3. Egg and sperm meet and fuse = fertilisation → cells divide to form a ball → 1 week – cells begin to specialise → 4 weeks spine and brain forming, heart beating → 9 weeks eyes and ears forming, movements start → 12 weeks fetus has most organs and moves – very tiny → remaining 28 weeks – fetus grows and matures → around 40 weeks baby is born.
4. The uterus provides the developing fetus with a warm, protected space to grow. It produces the placenta because the developing fetus needs a good supply of dissolved food and oxygen which is supplied by the placenta through the umbilical cord. It also needs its waste products to be removed through the placenta, and to be protected from infections and harmful substances. It needs to be supported and protected from physical harm by the fluid sac.

11.6
Health of the mother, health of the child

Student Book pages 204–205

kerboodle

Online resources

- Worksheet 11.6.1 From a fertilised egg to a baby
- Worksheet 11.6.2 The effect of caffeine on the growth of a fetus
- Worksheet 11.6.3 Extension: Thalidomide: a tragedy that changed the way medicines are developed
- Interactive test

Objective

- Discuss how fetal development is affected by the health of the mother, including the effects of diet, smoking and drugs

Overview

In this lesson, students apply what they learned in the previous lesson about the development of the fetus and the role of the placenta in supplying food and oxygen from the mother to the fetus to consider how the health of a mother, the diet available to her and the substances she chooses to take into her body may affect her unborn child.

Treat this topic with sensitivity as students' families may have been/be affected by some of the issues raised.

Activities

- Have a sketch on the board or a projection of a fetus in the uterus, e.g. Fig 11.5 B with no labels visible. Ask students to name the placenta, uterus, fetus, umbilical cord and fluid sac. Ask them to explain briefly how the fetus gets the dissolved food molecules and oxygen that it needs. Remind students that a fetus must get everything it needs from its mother. Introduce the idea that a woman must be careful when she is pregnant as what she takes into her body may affect her unborn child. She may make choices that help her child develop more healthily (see Unit 11.7) or events in her life may cause her growing baby harm.
- Read the section 'Food for mother, food for baby' with students. Discuss the kind of foods that a woman needs to provide for her growing fetus – such as iron for the blood, calcium for the bones, protein to make tissues and carbohydrates and fats for energy. Students answer questions 1 and 2 on page 205. They then use **Worksheet 11.6.1** to explore more about the rate of growth of the fetus during a healthy pregnancy and to consider the effects of poor nutrition during pregnancy. This is a Thinking and working scientifically activity involving data handling.
- Introduce the idea that it is not only food that crosses the placenta – drugs taken by the mother can also reach the fetus as it grows. Read the section headed 'Drugs and baby' on page 204 and the paragraph on medicines at the top of page 205. Explain clearly that the health of the mother is very important. If a woman is ill and needs medicine, then it is important that she gets it – but doctors must know that she is pregnant so they can prescribe her drugs that are safe for her developing baby.
- Ask students to name some recreational drugs. They may not know many, or they may have many different ideas. Without judgement, list them on the board and comment that many drugs cross the placenta and reach the developing fetus. Read through the section headed 'Recreational drugs' on page 205 with students. They then answer Q3a and 3b on page 205. Now give them **Worksheet 11.6.2** which provides them with a larger copy of the graph from Fig 11.6 D to use with Q3c on page 205 and an alternative activity to design a poster to explain to pregnant women why it is a good idea to reduce their caffeine intake during pregnancy.

Extension

Ask students to complete **Worksheet 11.6.3**. This is a challenging activity.

Homework

Workbook page 86

Key words

caffeine

11.6 Student Book answers

1. The fetus gets the nutrients its needs from its mother, through the placenta. Their blood supplies pass very close but do not mix.
2. She needs iron to make red blood cells to carry the oxygen she and her fetus both need, and she needs iron so her fetus can make its own red blood cells.
3. a. Medicinal drugs make you feel better when you are ill, recreational drugs do not; medical drugs are legal, many recreational drugs are not; people take recreational drugs because they enjoy them, they only take medical drugs when they are ill.
   b. Some medicines cross the placenta and reach the fetus. Some medicines stop the fetus from developing properly and may harm it. So doctors must be sure that any medicines they give to a pregnant woman are also safe for her baby.
   c. Fig 11.6 D shows that the more caffeine the mother takes in, the higher the risk of her having a low birthweight baby. Low birthweight babies are more likely to have problems or die. Every cup of coffee contains around 95 mg of caffeine, so by cutting back to just one or two cups of coffee a day your mother's friend could reduce her caffeine intake and reduce the risk to her baby. Best of all, cut out the coffee completely.

11.7 Understanding science, saving lives

SIC

Student book pages 206–207

kerboodle

Online resource

- Worksheet 11.7.1 Folic acid fortification and the rate of spina bifida per 10 000 births
- Interactive test

Objectives

- Discuss how fetal development is affected by the diet of the mother
- Evaluate issues which involve or require scientific understanding
- Describe how people develop and use scientific understanding both as individuals and through collaboration

Overview

This lesson helps students to develop their understanding of Science in context. It builds on work from the last lesson, the impact of the mother's diet on the development of her fetus, and also links back to work on the importance of vitamins in the diet from Stage 8, Unit 6.4. This lesson provides clear context for both areas of biology. It also provides opportunities for students to see how scientists and doctors have collaborated to share an understanding of one factor that reduces the risk of a fetus developing spina bifida, and highlights how understanding the science can empower women to make choices that will improve the chances of their offspring.

Deal with this topic with sensitivity as students or their families may have been/be affected by spina bifida.

Activities

- Read the introductory paragraph to students. Then ask them to summarise what they know so far about how the fetus gets the nutrients it needs to grow healthily. Find out how much students remember about vitamins and whether they can recall examples of vitamins from Stage 8.
- Read through the second paragraph and look at Fig 11.7 A with students, explaining what spina bifida is – many students may not have met the condition before. Describe the effects of spina bifida with sensitivity – they range from minor disabilities to being incompatible with life.

Extension

Students produce an information leaflet aimed at young men and women that explains the importance of folic acid in the prevention of spina bifida and the value of managing your diet and taking vitamin supplements before starting a family.

Homework

Workbook page 87

- Explain that for centuries people had no idea of the causes of spina bifida and we still do not know everything. Read the first paragraph headed 'Folic acid deficiencies' on page 206 with students. Ask what a peer reviewed paper means – make sure they all understand this.
- Discuss the ways in which scientific knowledge can be used by individuals, by medical professionals and by societies. Raise the issue of access – women cannot choose to eat folic acid-rich foods or take supplements if they do not know it is important or if they cannot afford/have access to the right foods. Students read the three bullet points – and then answer questions 1 and 2 on page 207.
- Give students **Worksheet 11.7.1**. Students then complete Q4 on page 207.
- Bring students together for a plenary session emphasising the importance of scientific understanding in decision making for both individuals and societies.

11.7 Student Book answers

1. a. A condition where the spine and spinal cord do not form properly in the fetus. It may be mild or very serious, causing paralysis.

b. Scientists have discovered that women who lack folic acid in their diet have a higher risk of having a baby affected by spina bifida, and that if women take folic acid before and during pregnancy, the risk of having an affected baby is greatly reduced. So it is important for women to have folic acid to reduce the risk of having a baby affected by spina bifida.

2. Individual women can choose to eat folic acid-rich food or take folic acid supplements before they become pregnant and while they are pregnant, to reduce the risk of having a baby affected by spina bifida.

Societies can choose to give all young women/pregnant women folic acid supplements OR to fortify staple foods such as bread and cereals so everyone has a good level of folic acid. This makes sure all women have plenty of folic acid whenever they get pregnant.

3.

Country	% change in the incidence of spina bifida since fortification of food
Canada	−54
Chile	−50
Costa Rica	−35
Oman	−80
South Africa	−30
US	−35

4. a.

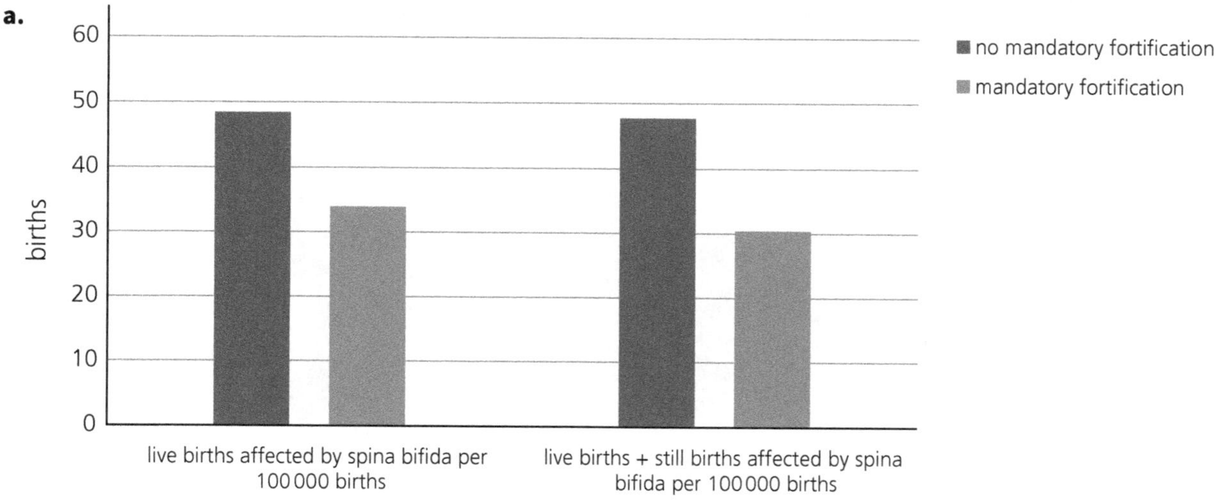

b. Look for students linking scientific understanding with changes in behaviour and positive health outcomes.

11.8 Smoking and pregnancy: the evidence

TWS

Student Book pages 208–209

kerboodle

Online resources

- Worksheet 11.8.1 Aleena's family history of births over 5 years
- Worksheet 11.8.2 Relative risk of stillbirth in mothers globally
- Worksheet 11.8.3 Extension: Risk analysis
- Interactive test

Extension

Students work through **Worksheet 11.8.3** which compares the risks of damaging a developing fetus from smoking with that of inhaling second-hand tobacco smoke.

Homework

Workbook page 88

Objectives

- Discuss how fetal development is affected if the mother smokes
- Make conclusions by interpreting results
- Evaluate the strength of evidence collected

Overview

This lesson covers the impact of smoking on fetal development, and provides the opportunity to develop aspects of Thinking and working scientifically – in particular, evaluating the strength of evidence collected by comparing personal small-scale data with large-scale studies. It also reminds students of their Stage 8 work on smoking and health. This lesson provides good opportunities for data handling.

Treat this topic with sensitivity as students' families may have been/be affected by some of the issues raised.

Activities

- Quiz students on how substances that a mother takes into her body – food, drugs, etc. – can affect her fetus. Also ask how the fetus gets the oxygen it needs. Ask students how smoking can affect the health of an individual – remind them of their Stage 8 work on smoking and health.
- Read the first paragraph of 'A cigarette for baby?' on page 208 and remind students that some substances taken in by the mother may harm her developing fetus. Now read the second paragraph.
- Work through Aleena's family history with students. Discuss what is meant by a cleft palate and a stillbirth. Remind students that babies born very small are more likely to have problems than average sized babies. Give students **Worksheet 11.8.1** which gives them a copy of the data table and squared paper for their graph.
- Read through Yousuf's online research with students. Highlight similarities and differences with Aleena's research into her family history. Focus particularly on the reliability of the data based on sample size – but also that Aleena's very personal experience mirrors global findings. Give students **Worksheet 11.8.2** which provides a copy of the data table and squared paper for their graph. Alternatively, students answer Q1a and 1b with graph paper provided instead of the worksheets.
- Students answer questions 2 and 3 on page 209.

11.8 Student Book answers

1. a.

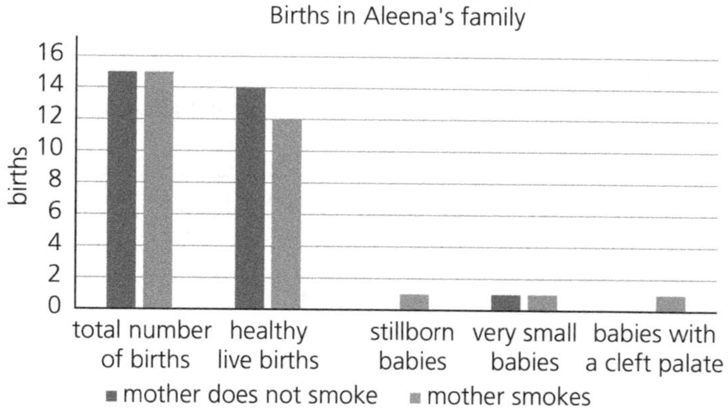

b.

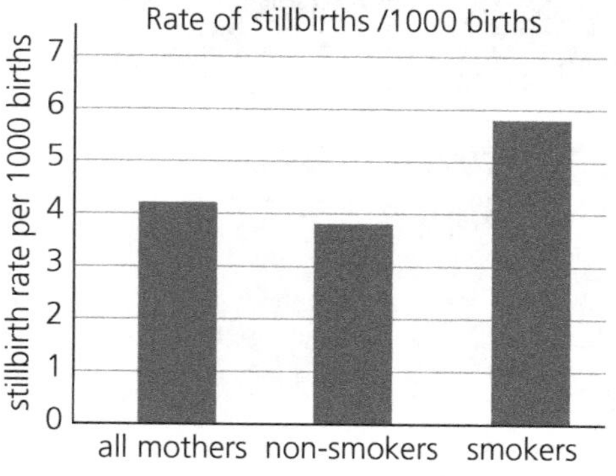

2. Evaluation should include:

Aleena: Negatives: very personal, very small sample, based only on what people tell her. Positives: very personal; two similar sized samples; results reflect those from larger studies; displayed data clearly; any other sensible points.

Yousuf: Negatives: Doesn't tell us where the information comes from; limited data. Positives: has used big, peer reviewed studies; looked for data based on samples of many thousands of women; found more detailed information about the risk of stillbirths; compared two studies; displayed data clearly; any other sensible points.

3. Main points:

- Smoking during pregnancy risks harming the fetus.
- Smoking during pregnancy increases the risk of low birthweight babies.
- Smoking during pregnancy increases the risk of the baby being born dead, both because it dies before it is born or because it dies during the delivery.
- Smoking during pregnancy increases the risk of the baby failing to develop properly e.g. it increases the risk of cleft palate and heart defects.
- Even if the mother does not smoke, if her husband smokes or she spends a lot of time in a smoky atmosphere, the baby is also at increased risk of low birthweight, early birth, stillbirth, cleft palate, etc.
- Any other sensible, thoughtful points.

11.9 Review answers

Student Book pages 36–37

Student book answers

1. **a.** A large molecule that acts as the genetic material in the nucleus of a cell. It carries the instructions to make a new cell/new life. **[2]**
   b. A long thread of DNA formed before a cell divides. **[2]**
   c. The section of DNA in the chromosome carrying information about particular characteristics. **[2]**
   d.

	Asexual reproduction	Sexual reproduction
How many parent cells are involved?	1	2
Are the offspring identical to their parents?	yes	no

2. **a.** cell X = sperm, cell Y = egg **b.** Gamete has half as many chromosomes as a normal cell. **[2, 1]**
   c. Tail = A; Cytoplasm with fat stores = B; Nucleus on both = C; Head/streamlined head = D; Jelly layer = E **[5]**
   d. A Tail = lets sperm swim to the egg; B Cytoplasm with fat stores = allows fertilised egg to grow and divide/produce an embryo; C Nucleus = contains the genetic material from the mother/father/ half the chromosomes; D Head/streamlined head = helps swimming/ breaks into egg; E Jelly layer = protects egg cell/attracts sperm. **[6]**
   e. Similarity – both contain half the number of chromosomes of normal body cells. Differences – any two from: egg much bigger than sperm; sperm can swim, egg can't move itself; egg female, sperm male; **[3]**
3. **a.** Fertilisation **b.** C E B D A **c. i.** 23 **ii.** 23 **[1, 5, 2]**
4. **a.** Your blood group is directly inherited from your parents – nothing environmental can change it. The graph shows there are 4 distinct possibilities with no overlap. **[3]**
   b. Identical twins are genetically identical so characteristics that are completely inherited, e.g. the shape of the earlobes and the blood group, will be identical whatever environment the twins grow up in. Height and body mass are partly inherited but they are affected by the environment the boys grew up in. If one twin had more and better food than the other, he will be taller and heavier. **[5]**
5. **a.** The pair of chromosomes that determine if you are female or male. **[1]**
   b. See Figure 11.3D page 199. **[8]**
6. **a. i.** = B **ii.** = C **iii.** = D **iv.** = A **[4]**
   b. Transports waste products from fetus to placenta/transports food and oxygen from placenta to fetus. **[2]**
   c. Supports the fetus as it grows and moves and protects it from external knocks and bumps. **[2]**
7. **a.** A **b.** B **c.** B **d.** A **[4]**
8. **a.** Nutrients pass from the mother to her growing fetus through the placenta. During the whole pregnancy, the fetus needs a balance of nutrients for its tissues and organs to develop properly. The mother must eat a healthy diet to provide her fetus with everything it needs. If the mother is deficient in a nutrient, her fetus will be also and this can affect its development. **[4]**
   b. During the last three months of pregnancy the baby grows and gains the weight it needs to survive after birth. It gets a layer of fat under the skin. The mother needs to eat a bit more to supply the baby with what it needs and maintain her own body. **[3]**
   c. Some medicines can cross the placenta and reach the developing fetus. Some medicines can harm the fetus and stop it developing properly especially in the first twelve weeks of pregnancy. Doctors need to choose medicines which will make the mother better but will not harm her fetus. **[3]**
   d. i. The mass of a baby increases as the pregnancy goes on. **[1]**
   ii. The evidence does support the idea that cigarette smoking can harm a growing baby. It shows clearly that the mass of a baby in the last few weeks of pregnancy is lower when mothers smoke than when they don't. Babies with lower birthweight are more likely to have health problems or die. **[3]**
   iii. Cigarette smoke contains many different chemicals which go into the blood of the mother. Carbon monoxide joins with the haemoglobin/red blood cells instead of oxygen, so a mother who smokes provides her developing baby with less oxygen for the developing cells. Many of the other chemicals in tobacco smoke cross the placenta from the mother to the baby. Many of them are toxins, so they slow down or affect the growth and development of the fetus. **[5]**

12.1 The carbon cycle

Student Book pages 212–213

kerboodle

Online resource

- Worksheet 12.1.1 Building a carbon cycle
- Interactive test

Prior learning

- Know that food chains can be used to represent feeding relationships
- Know that food chains begin with a plant

Extension

Ask students to produce a table comparing the carbon cycle in nature with the water cycle they learned about in Stage 7, Chapter 4.

Homework

Workbook page

Key words

carbon, carbon cycle, decomposition, combustion

Objective

- Describe the carbon cycle in nature

Overview

This chapter looks at the carbon cycle and climate change. In this first lesson, you introduce students to the carbon cycle, and the different processes by which carbon cycles through the living and non-living world. This topic provides opportunities for students to read and manipulate data. It also covers content which is important for success in both CLSS, and IGCSE and which will help your students become informed and scientifically aware global citizens.

Activities

- Ask students to sketch the water cycle they learned in Chapter 4. Then introduce the idea that water is not the only substance that cycles in nature. Read the first two paragraphs on page 212 with students, which explain the carbon cycle and why it is so important to living organisms.
- **EITHER** read through the paragraphs on respiration, photosynthesis, feeding, decomposition and combustion on pages 212 and 213 with students **OR** ask groups of students to make a brief presentation about each of these processes based on the content of the Student Book pages and anything else they know.
- Look at Fig 12.1 C showing the carbon cycle. Look at the interactions between different organisms and the atmosphere and discuss them with students, asking for examples of the different types of organisms involved.
- Give each student a large sheet of paper on which to create a poster using **Worksheet 12.1.1**.
- Students answer questions 1 and 2 on page 213. If they have not used the worksheet, they also answer Q3.
- Summarise the carbon cycle with students, making sure that they know the five main processes involved.

12.1 Student Book answers

1. a. A series of processes that move carbon between living things and the physical environment.

b. Molecules that make up living organisms contain carbon; plants need carbon dioxide for photosynthesis – important that there is always enough carbon.

2. a. Photosynthesis, respiration, combustion, decomposition, feeding.

b. Respiration: takes place in all living organisms. Glucose molecules are broken down using oxygen, producing carbon dioxide (CO_2) and water. This releases energy to be used by cells. The carbon dioxide is released into the atmosphere.

Photosynthesis: takes place in the green parts of plants, and in algae. Uses energy from light, trapped by chlorophyll in the chloroplasts. Carbon dioxide is taken in from the air and combined with water to produce glucose and oxygen. Takes carbon dioxide out of the atmosphere.

Feeding: Animals eat plants or other animals. Carbon is passed from organism to organism.

Decomposition: Microorganisms such as bacteria and fungi feed on the waste materials produced by animals, the dead leaves produced by plants and the dead bodies of animals and plants. They break them down, releasing carbon dioxide back into the atmosphere.

Combustion: Many fuels e.g. wood, coal, oil and gas, are carbon compounds. When they burn, they react with oxygen in the air, producing carbon dioxide and water which are released into the atmosphere.

3. Any diagram based on Fig 12.1 C.

12.2 People and the carbon cycle

SIC

Student Book pages 214–215

kerboodle

Online resources

- Worksheet 12.2.1 Sources of greenhouse gases
- Worksheet 12.2.2 Historical global carbon dioxide emissions
- Worksheet 12.2.3 Extension: Who produces most greenhouse gases?
- Interactive test

Prior learning

- Humans have positive and negative effects on the environment

Objectives

- Describe the impact of the carbon cycle on Earth
- Understand how using science can have a global environmental impact

Overview

This lesson continues directly from Unit 12.2 on the carbon cycle. Your students need to understand how people affect the carbon cycle and how that, in turn, has an impact on the climate of the Earth. They look at how human activities, often based on scientific discoveries, are adding to the greenhouse effect. One common misconception is that the greenhouse effect is a bad thing, when in fact it is the greenhouse effect that maintains a suitable temperature for life on the surface of the Earth. If you can support your students to reach this understanding, they will be well prepared for their further studies in IGCSE Biology. This lesson also contains many opportunities for students to practise their Thinking and working scientifically skills in handling and manipulating data.

Activities

- Sketch out a carbon cycle on the board by asking students to give you the different stages of the process.
- Read through 'A perfect planet' on page 214 down to Fig 12.2 A. Explain the greenhouse effect carefully, using Fig 12.2 A. Emphasise the importance of the greenhouse effect for life on Earth before moving on.
- Read through the final paragraph of this section – students should recognise that increases in carbon dioxide concentration in the atmosphere will lead to a rise in temperature at the surface of the Earth and that this will affect ecosystems.
- Work through the section headed 'People and carbon dioxide' with students and give them **Worksheet 12.2.1**. This supplies them with data for producing a pie chart showing some of the major sources of carbon dioxide resulting from human activities.
- Students answer questions 1 and 2 on page 215.
- Discuss the fact that different parts of the world produce different levels of carbon dioxide with students. Emphasise there are different ways of measuring carbon dioxide production, from total output of a country or region to the mean carbon dioxide produced per person. Students read the section headed 'Who produces most?' on page 215 before answering Q3. You can give students **Worksheet 12.2.2** to help them to produce the pie chart (Q3a).
- Summarise the main points of the lesson and ask students to consider the evidence they would look for ready for the next lesson.

Extension

Give students **Worksheet 12.2.3** to complete, as an example of a different set of data illustrating global differences in carbon dioxide production.

Homework

Workbook page 90

Key words

greenhouse gases, greenhouse effect, carbon sinks

12.2 Student Book answers

1. a. The greenhouse effect: Light from the sun reaches the Earth. Lots is reflected back into space. Greenhouse gases like carbon dioxide trap energy from the sun, keeping the surface of the Earth warm enough for life.
   b. Increase in carbon dioxide = increase in the energy trapped in the atmosphere = increase in temperature at the surface of the Earth.
2. Human transport systems: based on engines which burn fossil fuels like petrol; when fossil fuels burn they produce carbon dioxide which goes into the atmosphere. Humans use a lot of transport so a lot of carbon dioxide produced; fossil fuels like gas, coal and oil burned to generate electricity; carbon dioxide produced when they burn increasing carbon dioxide in the atmosphere.

   Deforestation: Removes trees that take carbon dioxide out of the atmosphere. The trees are burnt which adds carbon dioxide to the atmosphere.
3. a.

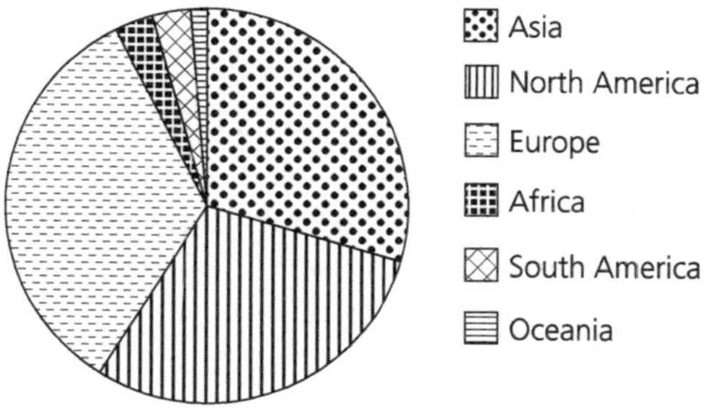

   b. Europe produces much more of the global carbon dioxide emissions, and Asia a lot less than in modern times; no data for international aviation and shipping.
   c. During most of the time period 1751–2017 there was no international aviation and shipping used wind power, not combustion of fossil fuels.
   d. Asia has grown more powerful/carries out many of the industrial processes/people everywhere travel more using vehicles/people everywhere want electricity, refrigeration, etc. so usage and thus generation and carbon dioxide production goes up – any other sensible points.

12.3 Historical impacts of climate change

Student Book pages 216–217

Objective

- Describe the historical impacts of climate change

Overview

It is a common misconception with students that climate change is something new. In this lesson, you show that climate change has happened both across the history of the Earth and in relatively recent history, highlighting the evidence that shows us these changes. This is another lesson that presents opportunities for developing TWS skills, as students are asked to manipulate and draw conclusions from data.

Activities

- Ask students what they know about climate change – vary your approach to the lesson depending on how well they understand the situation.
- Explain that the climate of the Earth appears to have been changing throughout history. Students read the section headed 'Climate change: ancient history'. Ask them to explain the different theories about cold periods and hot periods. Draw simple diagrams on the board to clarify the different causes. Students answer Q1 on page 216.
- Introduce the idea that the climate has been changing and getting warmer relatively quickly in recent history and that the changes seem to be linked to increases in greenhouse gases produced by human activities.
- Read through 'Climate change: recent history' with students and discuss the ideas, including whether there has been any evidence of climate change in your own region. Students answer Q2 to summarise this information.
- Give students **Worksheet 12.3.1** to help them answer Q3 on p217.
- Now give students **Worksheet 12.3.2**. They design a poster to warn of the potential problems that may arise as a result of global warming. Bring students together to summarise the learning points of the lesson.

kerboodle

Online resources

- Worksheet 12.3.1 Number of major named storms and hurricanes in 10 year periods
- Worksheet 12.3.2 The dangers of climate change
- Interactive test

Extension

Students examine Fig 12.3 C and write a paragraph to explain the impact of climate change on sea levels and why this is likely to have an effect on people around the world.

Homework

Workbook page 91

Key words

glaciers

12.3 Student Book answers

1. a. e.g. glaciation/global warming.
   b. Ash from volcanoes or dust from meteorite hits blocking the Sun; changes in orbit of Earth or activity of the Sun/increased carbon dioxide levels in the atmosphere.
2. Any four from: sea levels rising, flooding, drought, extreme weather events; Earth's surface temperature rising; ice melting.
3. a. Trend has been a steady increase in the number of major storms.
   b.

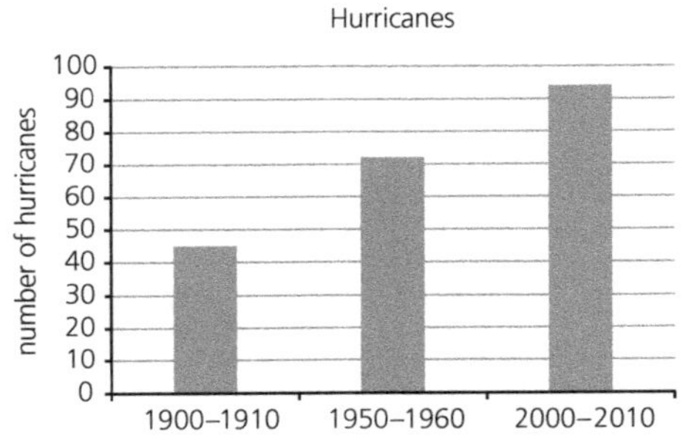

   c. The trend is similar – the numbers of major storms and the numbers of hurricanes increase steadily over time.
   Percentage increase in storms = 81/89 × 100 = 91%
   Percentage increase in hurricanes = 49/45 × 100 = 109%
   Rate of increase of hurricanes greater than the increase in tropical storms.
   Climate change is causing an increase in extreme weather.

12.4 Predicting the future

Student Book pages 218–219

kerboodle

Online resources

- Worksheet 12.4.1 Links and predictions
- Worksheet 12.4.2 Extension: Predicting the future
- Interactive test

Extension

Students complete **Worksheet 12.4.2**, another Thinking and working scientifically activity where they sketch graphs showing different scenarios if carbon dioxide levels increase, remain the same or fall and comment on the impact these different situations would have on conditions at the Earth's surface.

Homework

Workbook page 92

Key words

Intergovernmental Panel on Climate Change (IPCC), extinct

Objective

- Describe predicted future impacts of climate change

Overview

In this unit, students will learn how we try to predict the impact of future climate change and the difficulties of doing this. This lesson reinforces the potential impacts of global climate change and emphasises that these effects are not certain – they depend on decisions made by people and by the behaviour of each individual. There are opportunities for data handling and manipulation, giving students further valuable opportunities to practise the Thinking and working scientifically skills that are so important for success in IGCSE Biology.

Activities

- Remind students of the greenhouse effect – ask them to describe it.
- Read through the introduction and 'How hot will it get?' with students and explain the role of the IPCC as an organisation with representatives from many different countries collecting data from scientists all over the world. They use this information to assess what is happening and to predict what impact it may have. Look at Fig 12.4 A as an example of the type of data they collect. Read the first short paragraph of 'Future impacts' before students answer Q1 on page 219.
- Get different students to read aloud the paragraphs on the different predicted future impacts: 'Sea level rises', 'Droughts and heatwaves', 'Increased rainfall and flooding', 'Colder or warmer temperatures' and 'More extreme weather events'. Students answer Q2 on page 219.
- Read through the final paragraph 'Hope for the future' and discuss the difficulty of predicting the future. Ask students for reasons, e.g. people may change the way they generate electricity/develop new cleaner technologies/drive electric cars instead of burning fossil fuels/find ways to remove carbon dioxide from the atmosphere/plant more trees/stop cutting trees down/tackle the problems of getting all countries to work together/any other factor we haven't foreseen – encourage them to think of lots of ideas. Then give students **Worksheet 12.4.1**. This gives them bigger, clearer versions of the graphs in Fig 12.4 A to answer Q3 on page 219 – a Thinking and working scientifically activity, as they are interpreting and analysing data. It also provides a further activity which considers the difficulties of using evidence such as this to make predictions.
- Draw together the work so far on the difficulties of changing carbon dioxide levels or predicting climate change and global warming.

12.4 Student Book answers

1. a. Intergovernmental Panel on Climate Change.

b. It studies evidence from thousands of scentists all over the world and advises governments globally.

c. 1.3–1.8 °C

2. a. Sea levels will continue to rise: (26–82) cm by the end of the century.

b. Low lying countries will disappear, loss of coastal land (breeding grounds etc), etc.

3. a. 415 − 370 = 45 (allow ±2 ppm)

b. 0.6 °C

c. i. Carbon dioxide: between 2000 and 2020 (20 years) increase was 415 − 370 = 45 ppm. If rate continues the same for the next 80 years will be 45 × 4 = 180 ppm more than current level. 415 + 180 = 595 ppm in 2100.

ii. Surface temperature: between 2000 and 2020 (20 years) there was a 0.6 °C rise in temperature. If that continues for the next 80 years, the temperature increase will be 0.6 × 4 = 2.4 °C.

Allow appropriate calculations based on student's answer in **3a**.

Extension 12.5 Evaluating evidence for climate change

Student Book pages 220–221

kerboodle

Online resources

- Worksheet 12.5.1 Tree ring analysis
- Worksheet 12.5.2 Moving moth populations
- Interactive test

Objectives

- Describe the historical and future impacts of climate change
- Evaluate the strength of evidence collected
- Evaluate issues that require scientific understanding

Overview

This lesson may be used as an extension session but you can also use it for Thinking and working scientifically coverage, with lots of data analysis, and/or Science in context, with students looking at issues that affect everyone but that need scientific understanding to make the right choices. Students share some of the ways in which scientists gather evidence of historical climate changes and their effect on living organisms, and then use this evidence to predict patterns of change in the future. This work includes ideas that students may have met already, such as the analysis of ice cores, along with less well-known techniques such as tree ring analysis, bog cores and pollen species, and plant and animal distributions.

Activities

- Remind students that people all over the world are looking for ways to tackle rising carbon dioxide levels and climate change. This is all based on scientific evidence both of rising greenhouse gas levels in the atmosphere and changes in the global climate. Ask students how we evaluate scientific evidence and decide if it is reliable.
- Look at Fig 12.5 A on page 220 with students – they used this in the previous lesson as well. It is widely used as strong evidence for rising carbon dioxide levels in the atmosphere. Read the paragraph on the Mauna Loa data and ask students to list reasons why this data is widely judged to be reliable.
- Ask students to read through the sections headed 'Looking back in Time' on page 220 and 'Predicting the future' on page 221 and ask them to answer Q1 on page 221. Then go through and discuss their responses to the questions, highlighting strengths and weaknesses of the different types of data.
- Give students **Worksheet 12.5.1**. This makes it easier to answer Q2 as the rings are easier to see. Students are also asked to answer a further question about overall patterns in the climate during the life of this tree.
- Give students **Worksheet 12.5.2** to help them to answer Q3 on page 221. Ask them to answer Q3b by adding different coloured dots to the map on the worksheet, with a key to show what the colours represent.
- Look at Q4 on page 221 and have a class discussion on the importance of strong evidence. Remind students to revise for end of chapter review.

Extension

Students give a full written answer to Q4 on page 221.

Homework

Workbook page 93

12.5 Student Book answers

1. Mauna Loa curve: Accurate, reliable, good recording conditions/ only goes back to late 1950s, only taken in one place.

   Ice cores: give access to air bubbles thousands of years old, very pure/difficult to collect, accurate dating difficult.

   Tree rings: Clear record, can relate weather to evidence from other sources/have to cut the tree down, most trees don't live hundreds or thousands of years.

   Bog cores: clear evidence of climate at the time/ difficult to collect, depends on accurate identification of the plants.

   Changes in the distribution of plants and animals: modern evidence, collected now, clear changes, local impacts/ depends on accurate observations; difficult to tell if something is extinct.

2. **a.** 32 ± 2

   b. The weather was warm and wet. The tree grows well when conditions are warm and wet so it makes big rings like those seen here; it grows slowly when conditions are cold and dry so it makes small narrow rings.

3. **a.** 1968–1990 moths all clustered to the south of the country. 1991–2011 global temperatures increasing, moth moves north where it can survive as it is warmer.

   b. **i.** The distribution of the moth stays similar. **ii.** The moths will spread even further northwards.

4. Climate change driven by global warming is having many impacts, e.g. rising sea levels, increased flooding, droughts and extreme weather events. The evidence shows that the rise in global temperatures is the result of rising carbon dioxide levels in the atmosphere. The evidence is also strong that the rise in carbon dioxide levels is the result of human activities. This is important because people must change their behaviour to slow or stop the rise in carbon dioxide and so slow or stop climate change. It makes it easier to convince people when the evidence is strong.

12.6 Review answers

Student Book pages 222–223

Student Book answers

1. **a.** The carbon cycle is a series of processes that move carbon between living things and the physical environment. **[2]**

   b. **i.** B **ii.** C **iii.** D **iv.** E **[1, 1, 1, 1]**

2. **a.** Photosynthesis removes carbon dioxide from the atmosphere when it is absorbed by green plants. Photosynthesis uses energy from light, trapped by chlorophyll in the chloroplasts to combine carbon dioxide with water to produce glucose and oxygen used for respiration and to build the biomass of plants. **[3]**

   b. During respiration in animal cells, glucose molecules are broken down using oxygen, producing carbon dioxide (CO_2) and water. The carbon dioxide passes out of the animal into the atmosphere. **[3]**

   c. Wood and fossil fuels like coal, oil and gas contain carbon. In combustion, they react with the oxygen in the air to produce carbon dioxide and water. Combustion puts carbon dioxide into the atmosphere. **[3]**

   d. Decomposition is the process carried out by microorganisms such as bacteria and fungi that feed on the waste materials produced by animals, the dead leaves produced by plants and the dead bodies of animals and plants.
   They break them down, releasing carbon dioxide back into the atmosphere. **[3]**

3. **a.** Each segment is 18 degrees. Industry has 3 segments so $3 \times 18° = 54°$; $54°/360° \times 100 = 15\%$ **[4]**

b. Deforestation **[1]**

c. It reduces the numbers of trees and other plants. Plants remove carbon dioxide from the atmosphere in photosynthesis. Removing plants leaves more carbon dioxide in the air. Trees are often burned. Combustion of wood produces carbon dioxide raising the levels in the air. **[4]**

4. a. The concentration of carbon dioxide in the atmosphere in ppm (parts per million). **[2]**

b. Near the top of Mauna Loa volcano in Hawai'i. **c.** 1958 **[2, 1]**

d. The trend increases from 1958 to 2020. The rise in concentration is steady from 1958 to the 1990s. Then the rate at which the carbon dioxide levels go up increases. Any other sensible points, e.g. variations through each year. **[3]**

5. a.

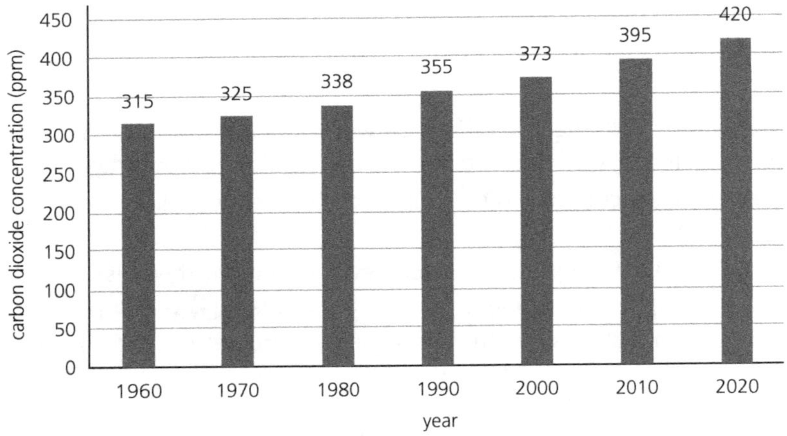

[5]

b. The concentration of carbon dioxide in the air increases, and the rate of increase gets faster in the last 30–40 years. **[3]**

c. Use of transport has increased, more electricity is generated by burning fossil fuels and the rate has accelerated, rate of deforestation has increased so more carbon dioxide being produced and less removed. **[4]**

6. a. Cooling; glaciation/warming. **[2]**

b. Ash from volcanoes, dust from meteorites, changes in orbit of the Earth or activity of the sun/ increased carbon dioxide concentration in the atmosphere. **[2]**

7. a. Intergovernmental Panel for Climate Change. **[1]**

b. Global warming is happening as a result of human activities such as burning fossil fuels and deforestation – these are causing an increase in carbon dioxide and other greenhouse gases in the atmosphere. **[2]**

c. Sea levels rising, increased flooding, increased heatwaves and droughts, more severe weather events. **[4]**

d. Sea levels rising: losing land, coastal erosion which can remove breeding areas for animals, loss of countries.

Increased flooding: loss of lives, loss of homes, loss of crops, topsoil washed away.

Increased heatwaves and droughts: crops failing, people and animals starving, extinctions, big fires.

More severe weather events: direct deaths, loss of infrastructure, associated floods, etc. **[8]**

8. a. Diagram similar to Fig 12.2.A. **[5]**

b. Greenhouse gases trap some of the energy from the Sun in the Earth's atmosphere. This keeps the surface of the Earth relatively warm so it is the right temperature for life. **[2]**

c. Current level of carbon dioxide in the atmosphere keeps the surface of the Earth at the right temperature for life. Increasing levels of carbon dioxide in the atmosphere will increase the amount of energy trapped and so increase the temperature at the Earth's surface. **[3]**

d. 1.3–1.8 °C **[2]**

e. Any selection of valid points based on different impacts of a rise in global temperatures on land, animals, plants and people. Look for descriptions, e.g. rising sea level, extreme weather events AND awareness of their impacts, e.g. loss of countries/land/breeding grounds. **[10]**

13.1 Variation in animals and plants

Student Book pages 224–225

kerboodle

Online resources

- Worksheet 13.1.1 Variation in animals
- Worksheet 13.1.2 Variation in plants
- Worksheet 13.1.3 Extension: Variation in the cabbage family
- Interactive test

Objectives

- Describe the variation in a species
- Relate this variation to genetic differences between individuals

Overview

This chapter covers variation, natural selection and extinction. In CLSS, students are NOT required to study genetics – what is important here is for students to understand that differences between the individual members of a species may result from genetic and environmental factors. This leads to an understanding of natural selection and factors that may lead to evolution. If students understand these broad principles in Stage 9, it gives them a firm foundation for later work on the mechanisms of genetics in IGCSE Biology. Key biological skills such as observation are also encouraged in this topic and are especially important in this first lesson.

Activities

- Begin by asking students questions to check how much they remember from their work in Chapter 11, e.g. ask: What is variation? What causes variation between individuals of the same species? Give an example of a human characteristic that is inherited. Give an example of a human characteristic that is the result of a combination of inherited characteristics and environment. Can you think of a human variation that would only be the result of the environment?
- **EITHER** read through Thinking and working scientifically page 224 with students **OR** ask students to read through this content themselves. Look carefully at the images of puffins in Fig 13.1 B. Ask students to identify similarities and differences between them and discuss how these differences might arise and why they are important. Students answer questions 1 and 2 on page 225.
- Students read the paragraph headed 'Human blood groups' on page 225 and then answer Q3 on page 225.
- Give students **Worksheet 13.1.1** and ask them to label the parts of the animal that might be affected by genetic variation, e.g. size of ears or eyes, hearing, number of offspring it produces, and the parts that might be changed by a combination of genetic and environmental variation, e.g. length of leg/height, body mass, length of tail, number of offspring that survive.
- Remind students that plants are living organisms too – and that variation is just as important in plants as it is in animals. Read through the final two paragraphs headed 'Variation in plants' with students. Look in detail at Fig 13.2 D which shows variation in a species of orchid. Students answer Q4 on page 225 and then look at **Worksheet 13.1.2**. Again, ask students to look at features of the plant that might show genetic variation, e.g. number of petals or anthers, basic leaf shape, or type of fruit or seed, and those that are affected by both genetics and environment, e.g. size of fruit and seeds, number of flowers, height of plant.
- Have a plenary session to summarise the lesson. Remind students that people have often made use of genetic variation to breed animals or plants.

Extension

Students use **Worksheet 13.1.3** as a basis for research on the way in which variation in different characteristics in the basic brassica plant has been used by people over many years to breed different types of crops.

Homework

Workbook page 94

13.1 Student Book answers

1. **a.** Genetic/inherited.
   b. Genetic/inherited; environmental.
2. **a.** Black and white/birds/with big colourful beaks/orange feet/live on rocky cliffs/have dark eyes and an eye stripe/beaks have red/blue/yellow stripes.
   b. Different sized beaks/different patterns of stripes on the beak/different depth of orange colour feet.
3. **a.** Genetic.
   b. There are four possible choices only/must be one of those types.
4. **a.** Three from: size of bulge at the bottom/size of flower/size of petals/shape of petals/depth of colour/shape of blue pattern.
   b. Some features are genetic: overall shape of the flower, pattern on the petals, colours.

   Environment only: lack of colour due to lack of minerals, any bits eaten by insects, any aspects of size or shape damaged by frost or heat.

   Combination: maximum size of plant/flower will be genetic BUT the size it reaches will be affected by the amount of photosynthesis that takes place which depends on amount of light and temperature. Appearance depends on the amount of water available.

13.2 Natural selection in action

Student Book pages 226–227

kerboodle

Online resources

- Worksheet 13.2.1 Describing natural selection
- Worksheet 13.2.2 Extension: Natural selection for survival
- Interactive test

Extension

Give students **Worksheet 13.2.2** where they look at the story of natural selection in the oysters of Prince Edward Island in Canada, draw a bar chart and explain the evidence in terms of natural selection.

Homework

Workbook page 95

Key words

natural selection

Objectives

- Describe the scientific theory of natural selection
- Relate natural selection to genetic changes over time

Overview

In this lesson, you introduce students to the key biological principle of natural selection and give them examples of natural selection in action. The fundamental idea is that most organisms produce many offspring and relatively few of them survive. Those that do survive will be the organisms with the genes that give them characteristics which are best fitted to their environment. These are the genes that will be passed on from one generation to the next. If you can help your students to develop a clear understanding of this idea now, then it will give them an advantage when they begin their IGCSE Biology studies.

Activities

- Remind students of the work on variation they carried out in their previous lesson.
- Read through page 226 with your students, explaining the process of natural selection clearly.
- Give students **Worksheet 13.2.1**. When they have completed ordering the sentences, go through the entire sequence with them. Students then stick down their sentences (or write them out for themselves), which gives them a clear description of natural selection to keep. If they decorate the page, it can help to fix the sequence in their minds.
- Now read through page 227 with students. Refer back to the pictures of puffins and orchids on pages 224 and 225 if needed. Discuss the different examples of natural selection given in Table 13.2.1 and ask students for any other examples of natural selection they can think of.
- Students work through questions 1–4 on page 227.
- Summarise the main points of the lesson. Introduce the idea that environment drives natural selection and ask students to think about examples of this and how it links to their work on the carbon cycle and climate change, before the next lesson.

13.2 Student Book answers

1. The parent organisms die, not all of the offspring survive.
2. a. A natural process by which the organisms best able to survive are the ones which live and reproduce, passing on advantageous characteristics to their offspring.
   b. Each individual inherits genetic variation from their parents and is different from all other members of their species. Individuals with characteristics that give them an advantage are the ones most likely to survive. The successful individuals survive long enough to reproduce and pass on their useful characteristics. This process is repeated many times until these characteristics become more common in the population. Over a long period of time, it may lead to the development of a new species.
3. Look for two local examples with clear understanding of how natural selection has led to particular characteristics, e.g. butterflies with long tongues to access nectar or similar.
4. Look for two local examples with clear understanding of how natural selection has led to particular characteristics, e.g. plants with deep roots to access water/flower at night for bat pollinators.

13.3 Environmental change and natural selection

Student Book pages 228–229

kerboodle

Online resources

- Worksheet 13.3.1 Studying natural selection in finches
- Worksheet 13.3.2 Extension: Charles Darwin and the theory of natural selection
- Interactive test

Objective

- Describe what happens to the population of a species when there is an environmental change

Overview

In the previous lesson, students learned that natural selection results in organisms that are well suited to their environment. In this lesson, you challenge them to think about the situation when that environment changes – this links holistically with the work they did in Chapter 12 on the carbon cycle and climate change. Making this sort of link has great value in helping students to see how the biology they learn fits together, and how each area of learning supports the others. Students who completed the Extension exercise for Unit 13.2 were introduced to environmental changes and natural selection, although they may not yet have made the link. If you did not use **Worksheet 13.2.2** in Unit 13.2, it also fits in well as an extra activity here.

Activities

- Question students to see how many examples of natural selection in action they remember from the previous lesson. Introduce the term adaptation – natural selection produces populations with adaptations that mean they survive well in their environment.
- Read through page 228 with students. Take time to discuss the ideas, take questions from students and help them to make the connection between environmental change and natural selection. Look carefully at the information in Fig 13.3 B, combining the written information with the data in the diagrams.
- Students answer questions 1 and 2 on page 229.
- Move on to read page 229 'Gradual change v. extreme events' with students, looking at Figs 13.3 C and 13.3 D before students answer Q3 on page 229.
- Ask students to complete **Worksheet 13.3.1**.
- Summarise the principle that populations change and become better adapted to a change in their environment as a result of natural selection, with students offering examples. If time allows, students can complete **Worksheet 13.2.2** if they did not do so last lesson.

Extension

Students complete **Worksheet 13.3.2**, which gives some examples of Darwin's work and other examples of natural selection.

Homework

Workbook page 96

Key words

adaptation

13.3 Student Book answers

1. If the environment changes, many organisms are no longer well adapted to their environment. Every population contains genetic variation. Some of the variants will be better adapted to the new conditions. They will breed successfully. As a result of natural selection most of the population will become this new form, adapted to the changed environment.
2. **a.** Best adapted; make the most seeds – 28 per plant.
   b. Both the 5-flowered plants and the 3-flowered plants cope better with the dry conditions than the 7-flowered plants. The 5-flowered plants make 15 seeds but the 3-flowered plants only make 6 seeds. The 5-flowered plants make more seeds so they reproduce more successfully/and become the most common form by natural selection.
   c. It would depend how the climate changes but one of the other variants might be better adapted to the new conditions. If it reproduces successfully it will become the main form by natural selection, e.g. if it gets wetter, the 7-flowered plant might become more common again by natural selection/if it gets even drier, the 3-flowered plants may cope better and be selected; even though they don't make many seeds they become the main form.
3. **a.** In normal years, whatever the temperature, kestrels raise an average of almost 3.3 young birds per nesting pair. When there is a drought, fewer young kestrels survive – on average fewer than 2.9 per nest.
   b. The data shows fractions of birds – but in any one nest it would be whole birds that survived. So must be the average of several or many nests in each case.
   c. As the temperature of the environment increases, those kestrels which are best adapted to the warmer conditions/e.g. eat a wider range of prey, regulate internal temperature better/will be more successful at raising chicks. Those less well adapted will raise fewer chicks – the overall effect will be that population numbers remain the same. In a severe drought it will be chance which birds find water or food – so population numbers fall and natural selection has little effect.

13.4 Extinction!

Student Book pages 230–231

kerboodle

Online resources

- Worksheet 13.4.1 Endangered animals
- Worksheet 13.4.2 Extinction rates
- Interactive test

Objective

- Describe how populations and species become extinct as a result of environmental change

Overview

In this lesson on extinction, students learn that species cannot always respond to changes in their environment and may become extinct. It is a common misconception that extinction is always a bad thing. In fact, 99% of all the species that have ever lived have become extinct – this is why we don't have dinosaurs like *T. rex* any more! Without extinction there is no room for new species. So, in biological terms, extinction is needed and important. The problem at the moment is the rate at which species are becoming extinct – it is too fast to be sustainable. These are the key concepts for this lesson. Students look at how people are having an impact and causing extinction. In Unit 13.6, they consider how people are also working to prevent extinctions. It is important that students see people as part of the solution as well as part of the problem.

Activities

- Begin by looking at Fig 13.4 A with students. Ask them about different types of dinosaurs and where the dinosaurs are now. Discover what they know about extinction.

- Read through 'What is extinction?' with your students, emphasising that extinction is a natural part of life when species cannot adapt to changes in conditions. Students read through the rest of pages 230 and 231. Divide your class into six small groups and give each group one of the endangered animals highlighted on **Worksheet 13.4.1**. Give students a few minutes to prepare, then each group describes their animal to the rest of the class and explains why it is endangered and at risk of becoming extinct. Discuss how these examples illustrate the information on pages 230 and 231. Collect and keep these cards for Unit 13.6.
- Students answer questions 1 and 2 on p231.
- Ask students to complete **Worksheet 13.4.2**. This provides them with a bigger, clearer version of Fig 13.4 D and an expanded version of Q3 on page 231.
- Have a plenary session on extinction and the threats to animals and plants caused by human activities and tell students that they will be looking at this further in another lesson.

Extension

Students write a report with the title: *Extinction: when natural selection fails.* Challenge them to explain extinction in terms of natural selection and to highlight the rapidly increasing rate of extinctions seen in the 20th and 21st centuries.

Homework

Workbook page 97

Key words

extinct

13.4 Student Book answers

1. a. Extinction is when there are no more individuals of a particular species alive either in an area (local extinction) or anywhere on Earth (global extinction).

b. Species become extinct because new species which are better adapted to the changed environmental conditions take their place. Extinction means there are not enough resources to go round.

2. a. Any four from: climate change; temperature of an area increasing; temperature decreasing; new disease; new predators; new competitors; changes in land use; catastrophic events; hunting; habitat destruction; pollution; any other sensible point.

b. Extinction happens because there are no longer enough organisms left to breed and maintain a population so that when the final organisms die out the species becomes extinct.

3. a. 1800

b. Amphibians. Global warming/loss of habitat; new fungal disease.

c. Data shows that extinction rates have increased rapidly over the last 200 years for all types of vertebrates. This is the period of time when we know the environment has been warming much more quickly. Suggests that the increase in extinctions is linked to the increase in global temperatures. Numbers increased particularly during 20th century when measured global warming also increased. Data show the percentage of all the species of each type of vertebrate that have gone extinct. There are thousands of species, so 2% of that is many species. If vertebrates are going extinct, invertebrates and plants will also be affected. There are many more species of invertebrates and plants than vertebrates so the numbers of extinctions linked to climate change is very large.

13.5 Investigating the peppered moth: past and present

TWS

Student Book pages 232–233

kerboodle

Online resources

- Worksheet 13.5.1 The colours of peppered moths over time
- Worksheet 13.5.2 Investigating Kettlewell
- Interactive test

Objectives

- Describe a model for natural selection
- Describe examples where scientists' unexpected results have led to improved scientific understanding
- Describe how people develop and use scientific understanding

Overview

This lesson helps students to develop their skills in Thinking and working scientifically.

It presents students with the story of the original work on peppered moths, with some of the evidence to support the theory. It contains plenty of data for students to interpret and analyse. They see how a changing environment leads to changes in the appearance of the moths over a relatively rapid timescale. Explain how scientists constantly question and challenge accepted ideas, to make sure that the evidence is strong. Finally, present modern work on the theory of the peppered moths, giving students the opportunity to analyse the data to see how it supports the original conclusions.

Activities

- Remind students of what they know about natural selection. Talk about timescales – originally people thought that natural selection always took a very long time.
- Introduce students to the peppered moth – they may be local to you as they are found in many countries around the world. Look at Fig 13.5 A on page 232 or project an image for everyone to look at.
- Read through 'Kettlewell's theory' with students. They should read on through 'Kettlewell's findings', looking closely at Fig 13.5 B and completing Q1 on page 233.
- Give students **Worksheet 13.5.1**. Read through the first paragraph on page 233 with students before they answer Q2 on page 233, then discuss their answers. Students continue to work through Worksheet 13.5.1, manipulating the data to show the changes in colour of the peppered moths in a different way and evaluating the different ways of displaying data.
- Talk to students about the doubts raised on Kettlewell's studies. Explain that scientists often challenge each other's work and ideas – this is how the evidence is kept strong. Students read through the rest of page 233 and answer Q3. **Worksheet 13.5.2** provides students with a copy of the table of Majerus's results and paper on which to complete their bar chart to answer Q3a.
- Summarise the findings of the lesson with students by asking them to identify a model of natural selection, how unexpected results led to an increase in understanding, and how our scientific understanding of the impact of industrial pollution on natural selection in peppered moths has developed and been used.

Extension

Students calculate the percentages of the different types of moths that were NOT eaten and add these to their bar charts.

Homework

Workbook page 98

13.5 Student Book answers

1. **a.** If natural selection causes changes in moth populations, clean woodlands will contain mainly pale moths. Pale moths are more likely to survive and reproduce, because they are camouflaged against the tree trunks where they rest. Darker moths produced by natural variation will be seen and eaten by birds. In polluted woodlands, he predicted that dark trees would make it easy for birds to find pale moths, so moths with genes for lighter colours would be eaten. As a result, in polluted woodlands there would be many more dark moths than light ones in the population.
   b. Looked at colours of old collections/took data from amateur moth collectors all over the country/did field experiments on birds eating different coloured moths.
   c. Old collections were a way of looking back at moths that had lived before the Industrial revolution. Taking data from moth collectors gave him evidence about colours of peppered moths in many populations and many different environments; field experiments allowed him to see if birds find more of one colour of moth than another on different coloured tree trunks.
2. **a.** Shows that the population of pale moths was very high until 1850. It fell until, by 1900, it was very low. Remained low until around 1970s when it increased again until, by around 2000, most moths are again the pale form. Shows that the population of dark moths was very low until 1850. It rose until, by 1900, it was very high. Remained high until around 1970s when it fell rapidly until, by around 2000, very few moths were the dark form.
   b. Until 1850, most peppered moths in clean woodlands were pale because dark form shows up clearly and those moths were eaten by birds so remain rare. From 1850s, woodlands became polluted as a result of industry. Became harder to see dark moths but easier to see pale moths so the proportions eaten by birds changed. After 50 years, mainly dark moths with just a few pale by natural selection. Around 1970s, pollution levels fell and woodlands became cleaner. Dark form became easier to see and were eaten; light form harder to see, so by the 2000s most peppered moths were again the light form by natural selection.
3. **a.**

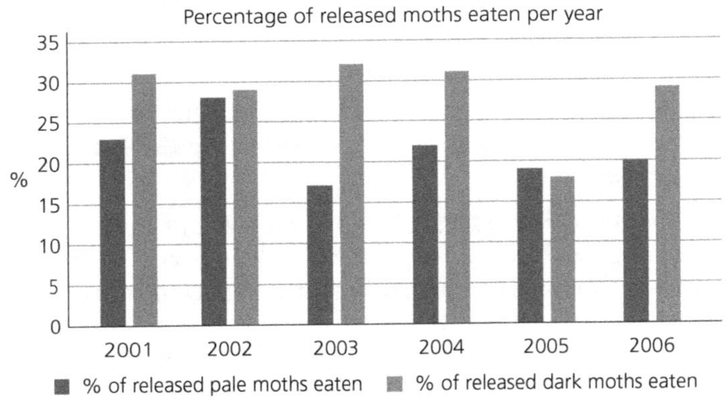

   b. Majerus carried out an investigation over many years. He observed peppered moths in their natural habitat to see where they rested and what ate them. He released 4864 moths over 6 years and observed where they rested, what ate them and the colours that were eaten. He built up a large data set which showed that, as Kettlewell observed, peppered moths do rest on tree trunks, they are eaten by birds and that, in a clean woodland, birds see the dark form more easily than the light and a greater proportion of the dark moth population is eaten. Thus natural selection will result in a larger number of pale moths as more of them live to reproduce. His findings confirmed the work of Kettlewell 50 years earlier.

13.6

What can we do?

SIC

Student Book pages 234–235

kerboodle

Online resources

- Worksheet 13.6.1 Helping endangered species
- Worksheet 13.6.2 Extension: Megafish v hydroelectric dams
- Interactive test

Objectives

- Describe what can happen to a species when there is an environmental change
- Describe how people develop and use scientific understanding
- Discuss how the uses of science can have a global impact

Overview

This lesson that helps students to develop their understanding of Science in context.

It shows students that people can be both the problem and the solution to environmental problems. They revisit their work in Unit 13.4 on extinction and see how human actions are helping to conserve all of the animals they studied. It is very important that students complete their CLSS Biology course with a sense of optimism – although humans have caused many problems, we also have the potential to solve them. The lesson also gives you the opportunity to highlight the idea of unintended consequences. This is a very important concept in environmental biology. Any change in an ecosystem – including changes designed to help particular species – may have unforeseen benefits or harms to other species. These ideas are presented to students through articles written by IGCSE Biology students – the next stage in their own studies.

Activities

- Ask a series of questions to remind students of the work they did in Unit 13.4 on extinction. Summarise the first paragraph of the spread, emphasising the point that people can both cause and solve problems. Split students into the groups that worked together in Unit 13.4. Give students their original animal cards from **Worksheet 13.4.1**, along with a new card from **Worksheet 13.6.1**. Give them a few minutes to prepare, then ask each group to present to the class a summary of the problems that are driving animals to extinction and then explain what actions have helped to reduce the threat to each species. Highlight the importance of ecotourism in placing value on animals and plants for local communities and providing a source of income, the importance of government involvement, the value of preventing habitat loss and the importance of captive breeding programmes around the world.
- Explain to students the problem of unintended consequences – how applying a scientific solution to one area can cause an unexpected problem elsewhere. Read through Article 1 with your students. Encourage discussion about the benefits of using hydroelectric power to produce clean electricity with no carbon dioxide. Consider hydroelectric power in the context of your own country – some regions already use a lot of hydroelectric power, others do not have the available water to do so. Encourage students to think about the suitability of hydroelectric power in different situations.
- Now read through Article 2 with students and again encourage discussion. Bring out the unintended consequences from producing clean electricity in this way.
- Students answer questions 1 and 2 on page 235.
- Briefly summarise the key ideas of this lesson and this topic before reminding students to revise for their review test next lesson.

Extension

Students complete **Worksheet 13.6.2**.

Homework

Workbook page 99

13.6 Student Book answers

1. a.

Advantages of hydroelectric power	Disadvantages of hydroelectric power
Relatively cheap	[No disadvantages are mentioned]
Doesn't burn fossil fuels so doesn't add carbon dioxide to the atmosphere	
Clean	
Plenty of rivers and water in many parts of the world	
Helps prevent global warming and climate change	

b.

Advantages of hydroelectric power	Disadvantages of hydroelectric power
Generates electricity	Drains water from areas
	Floods areas
	Prevents fish moving up and down river to reproduce
	Remove indigenous peoples
	Destroys ecosystems

c. Nearly one third of all freshwater fish species are threatened by extinction/16 freshwater fish species became globally extinct in one year in 2020/populations of migratory fish like salmon and eels have fallen by 75% in the last 50 years/megafish – enormous freshwater fish that weigh over 30 kg and live for many years are seriously threatened/Their populations have reduced by around by 94% since 1970.

2. a. Global warming and climate change are the result of increased greenhouse gases in the atmosphere. Generating electricity by burning fossil fuels adds carbon dioxide to the atmosphere/which increases global warming and climate change. BUT everyone wants access to electricity for all the benefits it brings. Hydroelectric power uses a natural renewable resource to produce electricity/and does not add greenhouse gases to the atmosphere/so helps reduce/prevent global warming and climate change./However, building dams traps water in huge reservoir./These flood the land./This may displace indigenous people/and destroys ecosystems./Changing the river flow can cause drought in other places./It stops fish from moving up- or down-river to breed.

b. To have the benefits without the problems means careful planning/and looking for both the advantages and disadvantages before starting./Make sure other areas or countries are not deprived of water./Build the dam where it will have least effect on indigenous people or ecosystems./Build ways for fish to make their way up and down stream around the dam./Any other sensible points.

13.7 Review answers

Student Book pages 236–237

Student Book answers

1. a. Differences between organisms of the same species. [1]
   b. Bar chart/graph [1]
   c. Inherited/genetic variation; environmental variation, e.g. different amounts of food given or found. [2]
2. a. O [1]
   b. AB [1]
   c. Body mass is decided by a combination of genetics and your environment. The amount you eat affects it. Blood groups are inherited/genetic variation. Nothing in the environment will change them. [4]
3. a. Natural selection is the process by which the organisms with the characteristics best adapted to their environment live and reproduce, passing on the useful characteristics to their offspring. [2]
   b. [5]

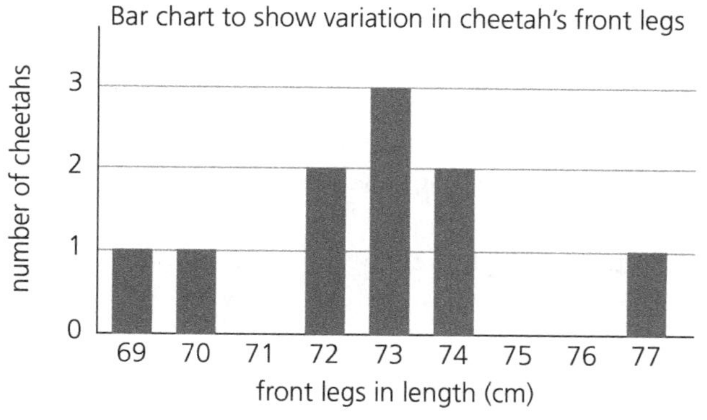

   c. A combination of genetic and environmental variation. [2]
   d. Cheetahs with the best leg length will be most able to run fast enough to capture prey without breaking their bones. These are the cheetahs that are most likely to survive to reproduce and pass on their genes for long legs. As a result of this natural selection, most female cheetahs have legs that are (72–74) cm long. A few cheetahs have short legs or particularly long legs as a result of natural variation. [6]
   e. Until the environment changed and a new plant-eater arrived, cheetahs with shorter legs were at a disadvantage, even though they used less energy. When a slower moving prey that reproduces quickly appears, shorter legged and slower cheetahs can catch them easily. As they also use less energy, they will be at an advantage, do well and become more likely to survive and pass on their genes.
   The population will change; most animals will have shorter legs and use less energy, as a result of natural selection. [6]
4. a. Organisms in a species show genetic variation → The organisms with the characteristics best adapted to the environment survive and reproduce. Less well adapted organisms die. → Genes from the successful organism are passed on to the offspring. They are likely to have the useful characteristics that made their parents successful → The process is repeated until most of the population have the useful characteristics. [6]
   b. A population or species is adapted to its environment. If the environment changes, e.g. warms due to climate change, the organisms will not be as well adapted. Some individuals in the population will be different as a result of natural variation. If some of the population cope better with the changed environment, they will be more likely to breed successfully and pass on their advantageous characteristics until most of the species have these characteristics and are adapted to the changed environment. [6]
5. a. [5]

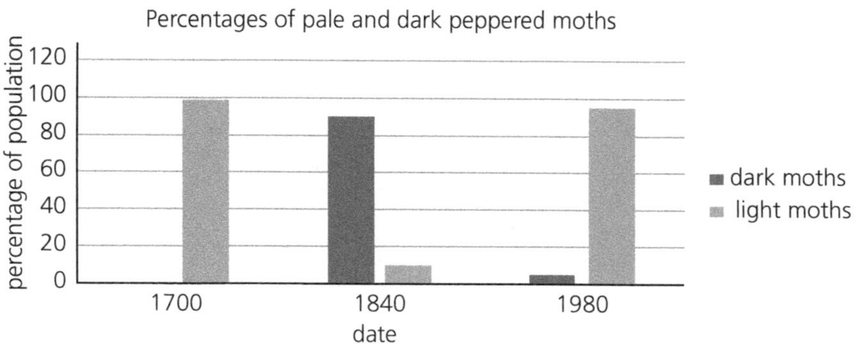

b. The population was almost entirely pale moths in 1700. By 1840, 90% of the population were dark moths. **[2]**

c. Pale moths would be more visible. **[1]**

d. Light moths became the most common form again, changing from 10% of the population to 95% of the population. The dark moths did the opposite and became rare again. **[3]**

e. In 1700, trees and buildings were clean. The pale form of the moth was hard to see. Dark moths were easy for birds to see and eat. Mainly pale moths survived to reproduce and so by natural selection most of the population was pale.

By 1840 the industrial revolution had resulted in a lot of smoke and many trees were blackened. Now the pale moths were easy for the birds to see and eat. Dark moths were camouflaged. They survived to reproduce and pass on the dark colour genes. The population became largely dark in colour by natural selection.

In the 1970s, the air was cleaned up. Trees became pale again. The paler moths had the advantage and survived to breed. Dark moths were seen and eaten. So natural selection drove the population back to being mainly pale moths with a few dark variations. **[12]**

6. a. Extinction is when there are no more individuals of a species left alive. It may be local or global, when an organism no longer exists anywhere on Earth. Any correct example, e.g. dinosaurs, dodo. **[4]**

b. The graph shows the human population growing steadily until the middle of the 20th century, when it starts to grow much faster. The rate of extinctions is very low and stable until around the time that the human population grows rapidly. Shortly after this, the numbers of extinctions rises very steeply and continues to do so. **[4]**

c. Once the human population starts to grow rapidly, human activities begin to change the environment and so drive organisms to extinction/any other sensible explanation. **[3]**

d. Any four sensible suggestions with reasons, e.g.:

Change the environment of an organism, e.g. global warming. Some species cannot cope with change in temperature and become extinct.

Change the environment so new diseases can take hold, wiping out species.

Introduce new organisms to an environment – may be new predators and wipe out species/or better competitors and wipe out species.

Loss of habitat causes extinction as species have nowhere to live or breed.

Hunting – guns and vehicles enable people to kill large numbers of animals and drive them to extinction/overfishing making fish species extinct. **[8]**

Stage 9

Review answers

Student Book answers

1. A = water, B = carbon dioxide, C = light, D = starch, E = oxygen **[5]**
2. **a.** The more carbon dioxide available to plants, the more biomass they produce. **[1]**
   b. **[3]**

$$\text{carbon dioxide} + \text{water} \xrightarrow[\text{chlorophyll}]{\text{light}} \text{glucose} + \text{oxygen}$$

   c. Carbon dioxide needed for photosynthesis. More carbon dioxide available to a plant = more photosynthesis. Plants make biomass by photosynthesis, so more photosynthesis = more biomass. **[4]**
   d. The seeds would respire and make carbon dioxide as they grew and any worms, etc. in the soil would also make carbon dioxide as they respired so the first box would have had some carbon dioxide in it, and the plants would have grown a little. **[2]**
   e. The amount of light – plants need light for photosynthesis – so they must all have the same amount for the results to be reliable.
   The amount of water in the pots – plants need water to grow and for photosynthesis so it's important they all had the same amount of water available.
   Temperature – the temperature affects how quickly plants photosynthesise so it's important they are all at the same temperature so they can be compared. **[6]**
3. **a.** Height of the seedlings/number of leaves. **[1]**
   b. Poor growth and the older leaves turn yellow. **[2]**
   c. Making proteins. **[1]**
   d. Leaves pale or yellow because magnesium needed to make chlorophyll which makes leaves green. **[3]**
4. **a.** Excretion is the removal of waste products from the cells. **[1]**
   b. Urea **[1]**
   c. A = kidneys which filter urea from the blood and produce urine
   B = ureter which carries urine from the kidney to the bladder
   C = bladder which stores urine
   D = urethra which carries urine from the bladder to the outside world. **[8]**
5. **a.** A = sperm; has a long tail for swimming/a streamlined head for swimming or breaking into the egg/genetic material in the nucleus to make a new cell. **[3]**
   b. B = egg; has food store to produce an embryo/genetic material in the nucleus to make a new cell. **[3]**
   c. The sperm breaks into the egg; the egg forms a protective layer so no other sperm can get in; the nucleus of the egg and that of the sperm fuse together to make a new cell. **[3]**
   d. **i.** X **ii.** X or Y **[1 + 1]**
   e. Diagram based on Fig 11.3 D. **[4]**
6. Diagram based on Fig 12.1 C – must show respiration, photosynthesis, feeding, death decomposition and combustion correctly linked but can be just words and arrows. **[11]**

7. a. Greenhouse effect: Sunlight reaches the earth. Some absorbed, lots reflected back. Some escapes Earth's atmosphere. Some trapped by greenhouse gases and reflected back to Earth.
Keeps the temperature at the surface of the Earth suitable/warm enough for life. **[5]**

b. Any two from: evidence from ice cores analysing the levels of carbon dioxide in the air – changes mean global warming or cooling.
Tree rings – get wider when temperatures are warmer.
Pollen grains from bogs show tropical plants growing/any other sensible suggestion. **[4]**

c. Human activities increasing the amount of carbon dioxide put into the atmosphere – through burning fossil fuels/agriculture, etc. AND reducing the amount of carbon dioxide removed from the atmosphere by cutting down rainforests. **[2]**
Overall increase in carbon dioxide in the atmosphere increasing greenhouse effect and raising temperatures art the surface of the Earth. **[2]**
Increased temperatures changing the climate so affecting animals and plants/more extreme weather events/melting ice and raising sea levels. **[2]**

8. a. Genetic/inherited variation; environmental variation **[2]**

b. Before 1950 a small number of rats in the population had genes that made them resistant to the poison, but it was not in use so this was not an advantage then. After the poison was introduced, many of the rats affected by the poison died. The rats with the genes which made them resistant had an advantage, so they were more likely to live and reproduce. They passed on their useful genes. By natural selection, the majority of the rats in the population ended up with the genes that made them resistant to the poison and the population grew again. As a result of variation, a few rats appear which are affected by the poison and as long as they don't eat the poison they remain in the population. Any other sensible points. **[8]**